MEASUREMENT

BOSTON STUDIES IN THE PHILOSOPHY OF SCIENCE

EDITED BY ROBERT S. COHEN AND MARX W. WARTOFSKY

VOLUME 72

KAREL BERKA

Institute for Philosophy and Sociology, Czechoslovak Academy of Sciences

MEASUREMENT

Its Concepts, Theories and Problems

Translated from the Czech by

Augustin Riska

D. REIDEL PUBLISHING COMPANY

DORDRECHT : HOLLAND / BOSTON : U.S.A.

LONDON : ENGLAND

Library of Congress Cataloging in Publication Data

CIP

Berka, Karel.
Measurement: its concepts, theories, and problems.

(Boston studies in the philosophy of science ; v. 72)
Translation of: Měření.
Bibliography: p.
Includes index.
1. Physical measurements. I. Title. II. Series.
Q174.B67 vol. 72 [QC39] 501s [530.1'6] 82-20428

ISBN-13: 978-94-009-7830-0 e-ISBN-13: 978-94-009-7828-7
DOI: 10.1007/978-94-009-7828-7

Published by D. Reidel Publishing Company,
P.O. Box 17, 3300 AA Dordrecht, Holland

Sold and distributed in the U.S.A. and Canada
by Kluwer Boston Inc.,
190 Old Derby Street, Hingham, MA 02043, U.S.A.

In all other countries, sold and distributed
by Kluwer Academic Publishers Group
P.O. Box 322, 3300 AH Dordrecht, Holland

D. Reidel Publishing Company is a member of the Kluwer Group

Original title *Měření: pojmy/teorie/problémy*

EDITORIAL PREFACE

For many years, Karel Berka has worked at some of the central problems of the theory of the sciences. At once a logician, a mathematician, a careful student of the physical sciences and the social sciences, and a sharp but sympathetic critic of the major philosophies of science in this century, Berka brings to this treatise on measurement both his technical mastery and his historical sensitivity. We appreciate his careful analysis of his predecessors, notably Helmholtz, Campbell, Hölder, Bridgman, Carnap, Hempel, and Stevens, and of his contemporaries such as Brian Ellis and also Patrick Suppes and J. L. Zinnes.

The issues to be clarified are familiar but still troubling: how to justify the conceptual transition from classification to a metric; how to explore ways to provide a quantitative understanding of a qualitative concept; indeed how to understand, and thereby control, the Galilean enthusiasm "to measure what is measurable and to try to render measurable what is not so as yet". Berka explores the vexed puzzle that arises when we explicate the so-called mapping of empirical relational systems onto numerical relational systems, empirical structures onto numerical structures: to what extent may we say that operations among the numbers tell us something about empirical operations, or other empirical relations? In fact, we see an epistemological and a historical discussion opening before us, as Berka proceeds to set forth the 'conceptual-mathematical' project of *metrization* in its several contexts of the empirical, operational, and theoretical practices of the *methods and procedures of measurement* in psychology, in physics and in all the intermediate scientific domains.

"What then is the object of measurement?" With this question, Berka initiates his searching examination of quantities, magnitudes, and numbers, and illuminates the conditions for extending about physical magnitudes to a generalized theory of magnitude, which in turn may allow us to formulate a clear and reliable theory of extra-physical magnitudes (and, in the process, to distinguish numbers from magnitudes, and magnitudes from quantities, and to sort out the epistemic nuances of scales, so easily construed as results of measurement but also as tools for measuring). Berka's critique of the additivity requirement, and of the proposal for an empirical operation of 'concatenation' (whether in physics or economics), seems to us the philosophical nucleus of his materialist theory of measurement, linked as closely as it is with the debate over extensive and intensive magnitudes in the measuring processes. But the reader will find discussions of philosophical importance throughout Berka's book, as in the deceptively clear account of imprecision in measurement, the extended methodological and epistemological analysis of operationism, the interesting, almost playful examination of the grave problem of meaningfulness and reliability in the use of measurements, the sharp judgment of any 'overestimation' of a conventionalist interpretation of 'measurements units'; and indeed in Berka's argued dismissal of any 'real' ontological basis for measurement which relies on a merely contemplative (which is to say, passive) materialist epistemology.

We may say that Berka brings us to the point of wishing another book, a fully developed theory of knowledge. For he bluntly concludes that measurement will be comprehended only within an analysis of its limitations as well as of its conditions and possibilities, and for him this must go beyond the phenomenalism, as he sees it, of Bridgman and even of the Carnap who is cited by Berka: " . . it is *we* who assign numbers to nature" (p. 215). Indeed, Berka concludes: "*some properties of real objects are not measurable in principle*". And then, with that in mind, how will the social sciences fare with respect to measurements and scales? We look forward to further reflections from Karel Berka's skilled and profound philosophical labors.

* * *

We are most grateful to Professor Augustin Riska for his careful and graceful translation, and also for the very thorough index of subjects. Our thanks too go to Susan Riska for the index of names, and to Carolyn Fawcett for her skillful editorial assistance throughout.

Center for Philosophy & History of Science
Boston University

ROBERT S. COHEN
MARX W. WARTOFSKY

October 1982

TABLE OF CONTENTS

TRANSLATOR'S REMARKS

(1) The Czech term *veličina*, occurring so frequently in the present work, has been consistently translated as *magnitude*, although in the contemporary English literature on measurement (for example, in Krantz *et al.* (1971) or Ellis (1966)) the term *quantity* is preferred. Yet the author needs the Czech term *kvantita* (translated as *quantity*) to be reserved for special philosophical purposes, as he explains in Section 3.1. Hence the use of the terms *quantity* and *quantities* instead of *magnitude* would conflict with the conceptual distinctions drawn by the author. At any rate, our use of the word *magnitude* is in accordance with the older usage, for instance with the terminology of Campbell (1920/1957), and also with the language adopted by Carnap (1966).

(2) Since *veličina* is being translated as *magnitude*, the Czech term *velikost* is translated as *size* (the size of a magnitude). This somewhat cumbersome term must be introduced in order to avoid formulations such as 'the magnitude of a magnitude'. Of course, if the term *quantity* could replace our term *magnitude*, one would get a more satisfactory rendering of the above phrase, namely 'the magnitude of a quantity'.

(3) The author distinguishes the terms *škála* and *stupnice* (see Section 4.1); as he expresses it, the term *škála* signifies conceptual scales, while the term *stupnice* signifies material scales. Hence, whereas the term *škála* is translated simply as *scale*, the other term, *stupnice*, is translated consistently as *gauge*.

(4) One has to be reminded of the important distinction between *numeral* and *numerical*; the adjective *numeral* pertains to the symbolic representation of numbers, whereas the adjective *numerical* refers to numbers themselves and their properties (the Czech equivalents are: *číslice, číslicový*, for numeral; and *číslo, číselný, numerický*, for *number* and *numerical*).

The terms *numerousness* and *numerosity* (following Stevens) are used in an alternative way, rendering the meaning of the Czech term *početnost* (in the sense of a number of . . . , an amount). Of course, *cardinality* can be used in this sense as well.

(5) The term *gnosiological* (occurring in the phrase 'ontologico-gnosiological') is not very common in the English literature, yet it is not replaced by the standard term *epistemological*, for it would then lose its special philosophical flavor. One may also take note of the frequently used phrase 'things, phenomena and processes of the objective reality'.

AUGUSTIN RISKA

PREFACE

The historical process of the development and utilization of quantitative methods in science, particularly in connection with the extension of measuring procedures in various domains of the social sciences, has conditioned a great upsurge of methodological investigations dealing with behavioral and social measurement. Starting with the contributions of S. S. Stevens toward the theory of measurement scales, problems of measurement have become the central theme of the methodology of empirical sciences, not only for the scientistic philosophy of science, but also for many psychologists, sociologists and economists. However, in a very extensive literature, which is focused on the practice and theory of measurement and scaling, two one-sided methodological positions are strongly exhibited: on the one hand, the position influenced by instrumentalism, operationalism and neopositivism; on the other hand, the viewpoint based on the formalistic philosophy of mathematics. From a purely empirical point of view, measurement is reduced only to the use of different scaling and measuring techniques. A purely mathematical doctrine of measurement, which prevails today, is reduced to the construction of various scales of measurement, defined merely by purely formally invariant transformations under which their form is unchanged, or, alternatively, to the derivation of the representation and uniqueness theorems from axiomatically defined relational structures. A common denominator of both these antagonistic doctrines is a very broad explication of the concept of measurement, encompassing a mere numbering, as well as an uncritical application of measuring procedures to the widest possible extent. Such explication disregards the general, objectively existing conditions of measurability, as well as the specific nature of the social sciences with respect to the natural sciences, especially physics.

In the present publication, which is the result of extensive work connected with the theory of measurement, we strive to analyze the problems of measurement scales on the basis of the methodological principles of dialectical and historical materialism. In this we start from the evident fact that measuring procedures have developed in the process of a practical appropriation of the world by man. From its very origin, measurement and its quantitative results have constituted a complex of mutually conditioned empirical and mathematical aspects, reflecting the dialectical unity of qualitative and quantitative characteristics of objectively existing objects, phenomena, and processes. The conceptual reflection and theoretical foundation of these mutual connections must fully respect not only the ontologico-gnosiological ground of measurement, but also the concrete conditions under which this empirico-mathematical method may be applied in various disciplines. The boundaries between what can and cannot be measured, are therefore, in our view, marked out contingently upon the historical level of our knowledge of the laws of the objective reality. Hence, these boundaries cannot be changed in an arbitrary manner. Certainly, we encounter many problems in various doctrines of behavioral and social measurement as well as in the theories of scales which are tied up with these doctrines. Nevertheless, these problems definitely cannot be short-circuited by a conventional choice of scaling techniques that are not theoretically justified, or by a construction of scales which cannot be empirically realized in a satisfactory way, or, finally, by formal means, however refined, with the help of which one cannot in fact solve any problem of content in a meritorious manner. An unfounded utilization of mathematical methods has a harmful rather than a beneficial effect; in other words, such utilization diverts us from the elucidation of problems which deserve priority from the theoretical point of view, and puts an additional burden on the solution of questions which are a necessary precondition of appropriate quantification.

It is exactly this standpoint from which one is to understand our discussion of the concepts of the origin of a scale, distance, scale types and transformation criteria. Yet, should the reader of our exposition get the impression that by our critical objections against the 'wider' doctrine of measurement we wished in principle to cast doubt on the possibility of social measurement, such impression would amount to sheer misunderstanding.

KAREL BERKA

1. INTRODUCTION

Comenius:
"In the number, weight and measure of things great secrets are hidden" (*Geometry*, Introduction).

Measurement is a quite common procedure in the everyday practice of humankind and each individual. Measurement and counting – operations directly connected with the material life of society – have developed in the process of the practical appropriation of the world by man. Both these methods have become an integral component of material production and distribution, as well as the objective point of departure for the rise of mathematical disciplines, especially geometry.[1] With the help of these disciplines, measurement and counting also provide a necessary precondition for the development of science and technology.

At the very beginning of their development, various ways of counting and measurement were obviously of an elementary nature: counting objects of a certain kind was based on a comparison with the number of fingers; measurement of distances was supported by comparisons with the length of the human thumb, foot, arm or stride. This simple way of quantification, which was "primarily and literally an experimental counting and measurement technique",[2] reaches far back into antiquity. Entire centuries had to pass before well-elaborated and differentiated methods and techniques of measurement had been successively established in various scientific disciplines. This process was connected with the development of mathematics and other sciences and with the advancement of the technology of measurement which was enforced by the demands of production, trade, communication, and the like.

Ordinarily, when we use measurement procedures, we are not at all aware of this complex historical development. Weighing trade articles on various kinds of balances, measuring time by using clocks, measuring the speed of a car with the help of an odometer – these are operations which are performed by contemporary man in a mechanical manner. These operations appear very easy and obvious. In order to find out the time, weight or speed of something, to be able to say how heavy it is, what time it is, or how quickly we drive, is it not sufficient only to watch the pointer on the scale of the respective measuring device? This way of measurement by *reading* the data off the scales of measurement tools or instruments seems so easy only because we do not consider all the necessary empirical and theoretical presuppositions which facilitated the construction and employment of these measuring devices.

Because of the experience, which we have acquired in the course of our own practice and which is justified by the bits of information we have learned at school, there will obviously be no problem when we weigh different objects by using a standard balance which has arms of equal length. For the purposes of the practical measurement of the weight of bodies – more precisely, their mass – it is sufficient to employ three rules which can be formulated, for the sake of simplicity, as follows:

(1) *The rule of equality*. If we place one body on one pan of a balance and another body on the other pan, these bodies will have the same weight if the pans remain in an equilibrium. If the pans do not remain in equilibrium, one body will be heavier (or lighter) than the other.

(2) *The additive rule*. If we place two bodies on one pan of a balance and weigh them (in accordance with the previous rule) as if they were one body, the total weight will be the arithmetical sum of the weights of both bodies.

(3) *The rule of the measurement unit*: When weighing, we place on one pan of a balance (which has arms of equal length) certain standard bodies, so-called standard weights, which are the prototypes of the basic unit of measurement,[3] as well as its multiples and portions.

With the exception of the third rule, we do not have to be explicitly aware of the previous two rules of weighing, since weighing is an obvious procedure in common practice. Even less would we consider the technical, physical, mathematical or conceptual conditions of this measurement procedure.

Also, we shall encounter no problems in simple measurements of distances, which are governed by three similar rules, expressible as follows:

(1) *The rule of equality*: If we place two bodies, which are bounded at both ends by straight line edges, longitudinally side by side (or on each other, or under each other) so that their edges coincide at one end, then these bodies will have the exact same length when their respective edges coincide with the other end as well. If they do not coincide, then one body is longer (shorter) than the other.

(2) *The additive rule*: If we place two bodies lengthwise, one after the other, so that their vertical edges are exactly contiguous, then the total length of this connected body will be the arithmetic sum of the respective lengths of both given bodies.

(3) *The rule of the measurement unit*: When we measure lengths, we use as a standard body some accepted length measuring device that is the prototype of the basic measurement unit of length – in our system of measurements, one meter – as well as its multiples or portions.

If we are interested only in finding out whether one body is shorter, longer or equally as long as the other body, or if we are concerned only about the ratio of two lengths, the first two rules will be quite sufficient. Of course, without the application of the third rule, we cannot say precisely how long the measured body is.

If we use all three rules, the whole process of measurement proceeds in a very simple manner. Let us assume that we are to measure the length of a beam. If we use a wooden *two-meter* as a measuring device, the comparison of its length with the length of the beam might lead to the following possible results:

(1) If the measured beam coincides exactly with our measuring tool, both bodies have the same length – in our case, two meters – and the process of measurement is finished.

(2) If the beam coincides with a portion of our measuring tool, the length of the beam is shorter than the length of the *two-meter*. The specific numerical value will be given by the coincidence of the second (rear) edge of the beam with some portion or multiple of the basic measurement unit which is marked on our measuring tool. Similarly, as in the previous case, the process of measurement will now be finished, and the additive rule was not employed at all.

(3) If our measuring tool coincides with a part of the beam, the measured object is evidently greater than two meters. Now it is necessary to repeat the entire process of measurement, by using the additive rule (which is related to the measuring tool) until, at the *n*th step, one of our two previous cases takes place.

The procedure of measuring lengths, which we have just described and exemplified, is also intuitively quite evident and simple, although it is open to a number of sceptical questions. Shall we always succeed in the repetitious placement of the two-meter along the measured object so precisely that we can utilize the arithmetical operation of addition? What if our measuring device gets bent? Shall we be able to read the corresponding numerical datum from it in an exact manner? How shall we determine the numerical value of the length of the measured object if it coincides with the measuring instrument off the marked sections of the basic unit of measurement, for instance, in between two-millimeter sections? What numerical values do we reach with this sort of length measurement? The answers to these and similar questions are already connected with the general theory of measurement, which will be treated later, or with the theory of length measurement, which is too specific and thus goes beyond the framework of our analysis.

For the sake of further argument, let us now mention only the answer to the last question, thus postponing a more detailed justification: numerical values acquired through a specific process of measurement are, at most, the values from the realm of rational numbers. This contention has important consequences for certain kinds of length measurement which are performed under similar empirical conditions by employing the same rules.

Let us imagine that we have at our disposal a certain body which has the shape of a right-angled isosceles triangle with the sides $a = b = 1$ m, and that we have to measure the length of its hypotenuse c. If we proceed in the same manner as in determining the length of a beam, we shall get as the results of measurement some numerical values from the realm of rational numbers, say, 1.42 or 1.41 m. Yet, if we consider the Pythagorean theorem, which claims that $c = \sqrt{(a^2 + b^2)}$ holds, we get, in our case, $c = \sqrt{2}$. As we know very well, this number is an irrational number which is mathematically expressed by an infinite non-periodical decimal expansion, namely 1. 411 213 5 . . . The length of that edge of the body which corresponds to the hypotenuse of the right-angled isosceles triangle ($a = b = 1$ m) cannot thus be measured by means of the above-mentioned method, even if we use a measuring instrument with arbitrarily small segments of the basic unit of measurement.

The conflict between these two very plain cases of length measurement shows very convincingly that the problems of measurement are not so simple as they appear at first glance. This statement can be very easily confirmed by a whole multitude of other cases. Just take, for instance, the measurement of temperature where we cannot use the additive rule, or the measurement of utility where the rule of the measurement unit is not met.

We know about the difficulties encountered by the Greek surveyors and mathematicians in their attempts to solve the problems of incommensurable lengths, as well as about the problems of measuring temperature with which physicists had struggled for a long time, or about the troubles with which psychologists, economists or sociologists now wrestle in the various areas of measurement outside physics. However, in none of the historical epochs could these difficulties prevent a very intensive search for ways of applying quantitative methods in the empirical sciences; on the contrary, they rather stimulated such an effort. In the process of knowledge, the quantitative approach – as well as the qualitative one which held the priority in the entire evolution of scientific cognition, for man had first to discern properties and things on the basis of similarity and difference – is objectively rooted in the very essence of objective reality, namely in the fact "that the relation of quality and quantity is reciprocal, that quality can become transformed into quantity just as much as quantity into quality, that, in fact, reciprocal action takes place."[4] At any rate, the significance of quantification in various spheres of human activity, irrespective of the obstacles which must be overcome practically and theoretically in the process of its application, springs, in a reflected form, quite directly from the knowledge of its advantage and utility.

The relation between the qualitative and quantitative approaches and their evaluation can be explicated on the theoretical level by the familiar division of concepts into classificational (qualitative), topological (comparative) and metrical (quantitative).[5]

Classificational concepts, such as *close, cold, long, old*, which are determined only in a qualitative sense, serve the classification of objects on the basis of common characteristics. In the evolution of scientific knowledge, these concepts are primary, for both historical and systematical reasons. They are characteristic of the stage of scientific *classification* through which every scientific discipline had to pass. Nobody can doubt the importance of these concepts from the viewpoint of taxonomy which is applied, for instance, in botany, zoology, mineralogy, and the like. Of course, we must not forget that these concepts are inexact and indeterminate, and that their use in the conceptual construction of scientific theories is considerably limited.

As each of us knows well from our own experience, communications such as *I shall come immediately, The nearest town is far away* and *Water is warm*, are very vague and their sense varies from individual to individual. If a member of a club of hardy people says that the water in Vltava is warm, the majority of people in Prague will judge that it is ice-cold. If we find out from a Siberian hunter that it is not far to the nearest town, we shall not view this distance in the same way as when we get similar information from a taxicab driver in Estonia.

Furthermore, classificational concepts can offer us only a small number of subclasses when we use them for the classification of a certain concept. In some cases, for instance, in the classification of basic colors, student achievements, etc., this fact gives us no real problems. However, when we consider classificational concepts which express properties capable of gradation, this limitation will already disturb us. For example, when expressing different states of temperature, we might use the following classificational concepts: *hot, very hot, warm, lukewarm, cold, cool* and *icy*. This classification could, of course, be made much finer. Nevertheless, even if we introduce additional categories, for instance, *very cold, very cool* or *awfully icy*, this would not be enough for a more refined classification, not to mention the fact that we might apparently encounter difficulties of interpretation if we try to distinguish the differences, say, between *very cold* and *cool*. In principle, we could, of course, introduce as many classificational concepts as we need specific terms for every really relevant degree of a given property. Apparently, our common stock of words would not suffice to cover this inefficient way of expression.

The meaning of some, but of course not all, classificational concepts can be made more precise by expressing them in a relational form, i.e., as topological concepts; for instance, . . . *warmer than* . . . , . . . *longer than* . . . , . . . *harder than* . . . , . . . *as old as* . . . . *Topological* concepts have already a greater cognitive value than classificational ones. They enable us, not only to establish the sameness (or difference), but also to mutually compare at least two objects which possess a given property and, consequently, to arrange them into a sequence.

If we know, for instance, that object X is longer than object Y and also that object Y is longer than object Z, we can immediately conclude from this that object X is longer than object Z. Of course, we may not be able to give specific lengths of objects ordered in this way, even if we know, for example, the numerical value of the length of object X. Only in an exceptional case, if we use an expression of the type "A is as old as B", and if we know by chance the age of one of the two persons, can we, on the basis of this concept, immediately determine the age of the other person.

It is characteristic of topological concepts that they provide a transition from classificational to metrical concepts. From the methodological standpoint, topological concepts have much more in common with classificational than with metrical concepts, for an ordering of objects with respect to a certain property might be regarded as a kind of classification which shows not only the relations of subordination, superiority and co-ordination between particular subclasses, but also the mutual arrangement of co-ordinated subclasses. Furthermore, one finds here a very fine classification which, if necessary, can in principle be made even more refined, namely by introducing a further *intermediary link* between any arbitrary members of the ordering. Of course, this procedure assumes that we are dealing with a property capable of gradation; for instance, in the case of Mohs' classification of hardness of minerals, which is based on the topological concept . . . *harder than* . . . . However, when we classify by means of properties which do not permit gradation, for instance, *Pilsner, feline, wooden*, etc., only classificational concepts can be applied.

From the conceptual standpoint, topological concepts are more closely connected with metrical ones, for the existence of a topological concept – more precisely, the satisfaction of the conditions of ordering (topologization) – is a necessary presupposition for the existence of a metrical concept.

Metrical concepts, such as "50°C warm", "10 m long", and the like, not only express a qualitative characteristic, for example, length, temperature, etc., but already give exact quantitative specifications. In this way, metrical concepts explicitly reflect the unity of quantitative and qualitative aspects of real objects. Their objectivization, of course, is relative in respect to the accepted system of the units of measurement. This specification is indeed necessary, for we might not otherwise know that the metrical concept "the state of temperature 122°F" has exactly the same meaning as the concept "the state of temperature 50°C".

The significance of metrical concepts does not rest primarily on the service they provide for a more precise interpretation of indeterminate and vague predications in statements of the type "Water is warm", with the help of quantitative statements, for instance, "Water is 27°C warm," which have an intersubjective character. The relevance of these concepts is based, first of all, on the fact that they enable us to formulate numerical laws which are inevitable components of all exact sciences, as well as to utilize mathematical means in the empirical sciences.

The transition from classificational to metrical concepts, which on a theoretical level expresses the measurability of the respective property, must be carefully distinguished from a more or less successful attempt to provide a *quantitative explication* of a certain qualitative concept. To give an illustration of this manner of conceptual elucidation, we may mention the exemplification of a certain qualitative concept by means of numerical data; for example, the explication of the concept of a successful theatrical production by the number of reruns, or the discovery of a suitable correlation between qualitative and quantifiable concepts which have a different content, for instance, between the concepts of fear and the adrenalin level in blood.

One can appreciate the importance of this differentiation if one considers the fact that a quantitative explication is often interpreted in a reductionist spirit and that this interpretation is then, without justification, transferred to the view of metrical concepts. The objective transition from classificational to metrical concepts, which is historically conditioned and which cannot be materialized without sufficient reasons, should never be conceived as a reduction of quality into quantity, or as a neglect of the qualitative aspects of reality. The incorrectness of such an interpretation must be evident to everybody who realizes that each metrical concept explicitly contains its qualitative specification. The legitimate interest in the advancement of quantitative methods in science has not only been conditioned in its historical evolution by sheer objective factors. Under the influence of the idealistic views of the Pythagorean-Platonic type, it is subjectively tied up with the idea of a universal mathematical method that could be generally applied in all domains of scientific knowledge.

The fact that this methodologically incorrect and philosophically untenable idea of a universal mathematical method has greatly influenced the development of science in various periods of scientific revolutions, is an expression of the dialectics of the historical process of knowledge. Nevertheless, this idea is based on a misconception of the process of abstraction and idealization, the absolutization of the quantitative aspects of reality and an overestimation of the importance of quantitative concepts in the process of knowledge. During the Renaissance, this idea influenced the development of astronomy, mechanics and, in step-by-step fashion, other natural sciences too. During the nineteenth century, it started to infiltrate psychology and economics and, in our century, it has influenced the development of sociology, pedagogy, linguistics and other social sciences.

In the science of antiquity, although this conception was programmatically expressed by Plato's slogan, "Do not enter unless you know geometry", it did not essentially succeed and so remained limited to the mathematical disciplines, among which one could also find astronomy or music. Because of the objective needs of classification and systematization of scientific knowledge, and owing to the influence of the scientific and philosophical views of Aristotle, antiquity may be characterized by the prevalence of a qualitative approach over a quantitative one.[6] In the Middle Ages, the confrontation between qualitative and quantitative approaches, which, from the standpoint of Marxist-Leninist philosophy, always one-sidedly emphasizes only one aspect of the dialectical process of knowledge, led to a temporary predominance of the qualitative approach, as a consequence of a deformed acceptance of the basic ideas of the Peripatetic school. And it is precisely this fact that is one of the essential reasons why Renaissance science, in its justified struggle against scholasticism, supported the effort to reach the true knowledge of nature (which was primarily conditioned by the advancement of crafts, trade and sea voyages), not only by a Democritean atomism, but also by the Pythagorean/Platonic ideal of mathematization.

The transition from the qualitative view of reality to a quantitative one is therefore regarded as the beginning of modern science, which found its embodiment in the work of Kepler, Galileo and especially Newton. Many historians of science claim that modern science arises from measurement and that it is inconceivable without it. Thanks to measurement, modern science has actually established itself. We do not consider this contention, which rests on Galileo's thought, "To measure what is measurable and to try to render measurable what is not so as yet",[7] as quite justified in its generality. Although it is true that, since the time of Galileo, the development of science has been connected with the use of mathematical methods and with the elaboration of methods of measurement, nevertheless, a merely one-sided application of quantitative methods and purely empirically-oriented measurements could never, in themselves, lead to the constitution of classical physics as represented in Newton's *Philosophiae naturalis principia mathematica*. In this work, just as in the works of Galileo, quantification is always subordinate to theory. To them, measurement is not the end of scientific knowledge, only its means.

The remarkable results which physics achieved by combining the experimental approach, based on measurement, with theoretical constructions, as well as by the natural connection between experience and theory that was mediated by mathematical means, have made a deep impression on other sciences and have become, especially now, the universal ideals of all sciences.

In the nineteenth century, this trend found two allies among the social sciences: economics and psychology. Under the influence of Bernoulli's hypothesis of the maximization of expected utility, which made a distinction between the value of money due to amount and the value of money due to utility for a certain individual,[8] bourgeois economics gave rise to a theory of utility in its two different variations. The adherents of so-called *ordinal* utility held the view that utilities can only be ordered; while the adherents of so-called *cardinal* utility considered utility as measurable in the strict sense of the word. In our century, von Neumann and Morgenstern built on these traditions of measurements in economics and unambiguously embraced the position of cardinal utility. Their views have crucially influenced current conceptions of measurement in the social sciences, particularly from a methodological standpoint. According to them, utility must be understood "as quantitatively measurable, i.e., as a number".[9]

The Bernoullian tradition had influenced the attitude toward the problems of *measurable* utility within a broader context of so-called subjective measurement, which encompasses the measurement of values, attitudes and other "subjective magnitudes", so intensively developed in psychology. Of course, the entire situation has been influenced by other factors as well, especially by the example of physics and by the philosophical doctrine of psycho-physical parallelism. The advancement of quantitative methodology in psychology, which was conditioned in the nineteenth century by splitting psychology into *empirical* and *philosophical*, has become the program of so-called psychophysics, which aims at the investigation of the relations between physical stimuli and psychical sensations on the basis of measuring physical magnitudes.[10]

In our century, under the influence of the philosophical doctrines of empiricism, pragmatism and, especially, operationalism, measurement has been elevated into the fundamental methodological postulate of empirical psychology, empirical sociology, empirical pedagogy, etc. Measurement has thus also become the theoretical foundation for the construction of streams or subdisciplines in various domains of the social sciences, as can be demonstrated by the titles of *econometrics, psychometrics* or *sociometrics.*

Of course, the paradigm of physical measurement, as well as mathematical method, cannot be applied in the same way in all scientific domains, for each scientific method is always tied up with specific problems of content and theoretical constructions which vary in different scientific areas. Therefore, the use of methods and conceptions taken from other scientific domains must always be appropriately modified, in order to safeguard a harmony between such applications, on the one hand, and the subject matter of the scientific discipline in question, as well as the degree reached of its theoretical construction, on the other hand. Measurement will lead to fruitful results in various areas only if there is a concern for the dialectical relationship between theory and practice, between the advancement of the methods and instruments of measurement and a specific theory of the measured magnitude, if we utilize measurement in productive or technical practice; or between testing hypotheses or verification of numerical laws and their theoretical justification, which depends on a general theory of measurement, if we carry out these operations within the framework of basic research.

The historical process of the advancement and use of methods of measurement, full of contradictions, also had to affect the development of theoretical and methodological conceptions and investigations which are related to the analysis and evaluation of this procedure.

In the theory and methodology of physical measurement, the entire situation was facilitated by the existence of the objective conditions of the applicability of quantitative methods, just as by the rich experience in the application of measurement procedures, which became the target of intense attention in the eighteenth century. Indeed, physicists did not feel the need to be preoccupied with the methodological questions of experiments, since they had not encountered serious obstacles in this respect. Their interest had been concentrated rather on questions of a practical nature: on the design of suitable and, as much as possible, more precise and advanced apparata that would enable them to extend the scope of measurable values of the given magnitude, to raise the measure of their precision and to facilitate control of the acquired numerical results by different methods of measurement. Since there were no serious doubts about the existence of various units of measurement, the question of their standardization was solved rather from a practical viewpoint. Although in our century this general attitude towards the theory and methodology of measurement has undergone changes, it has not essentially influenced the common theory and practice of physical measurement. With a few exceptions, measurement in physics is not loaded with unexplained or controversial problems of a theoretical and methodological character. Among these exceptions, one has to mention the discussion about the nature of measurement in quantum physics, in particular about the function of measuring apparata. One also has to mention the polemics concerning the nature of the operational concept of measurement, which quickly ended with unfavorable results for the position. In addition, mention should be made of the problems of relativity with regard to measurement, as well as of the theoretical considerations dealing with dimensional analysis, the practical application of which produces no problems, especially in technical physics.

Yet if psychologists, economists or sociologists attempt to follow the path of physics, they face problems of a quite essential character. Since they cannot consistently build on results that are verified by practice to a sufficient degree, or on a certified theory of extraphysical measurement, they try to justify the meaningfulness and realizability of measurement in their domains, at both the theoretical and methodological levels. At the same time, they try to repudiate any doubts concerning the legitimacy of quantification in these domains, when quantification is based on measurement procedures; a good example of such attempts can be found in the familiar discussion about the possibilities of measurement in psychology.[11] In both cases, these social scientists must defend themselves against two contrary, but in their consequences actually very close, views that are directly tied up with the evaluation of the applicability of mathematical methods in the social sciences. In their own camp, these scientists have to meet sceptical views concerning the *limits* of measurement in the social sciences. Such views spring from an overestimation of the qualitative character of these disciplines and from an underestimation of the advantages that follow from the suitable applications of quantitative methods. On the other hand, the social scientists must defend their own position concerning the *possibilities* of measurement in social scientific disciplines against natural scientists and engineers. Their own position is based, according to these critics, on an overestimation of the role of mathematical methods in the social sciences and on an underestimation of the specific content of the social sciences.

Toward the end of the nineteenth century and at the beginning of this century, mathematicians, too, indirectly paid attention to the theoretical questions of measurement in connection with the discussion about the relation between magnitudes and numbers, the nature of geometry and axiomatizations of the theory of quantities.

These different sources and areas of theoretical and methodological attitudes toward the problems of measurement obviously also affected the development and orientation of theoretico-methodological investigations in which we might distinguish roughly three periods. In our further outline, we shall restrict ourselves only to fundamental works – to those which have, in their own time or perhaps later, essentially influenced the contemporary theory and methodology of measurement.

In the first period of the investigation of theoretical and methodological questions of measurement, emphasis was put on philosophical and mathematical aspects. The essay of Helmholtz, *Zählen und Messen erkenntnistheoretisch betrachtet,*[12] which appeared in 1887, is generally regarded as a pioneering contribution towards measurement theory. This essay is devoted to the epistemological and mathematical aspects of quantification. Hölder (1901), Huntington (1902) and Wiener (1920) were concerned with the axiomatization of quantities, but only Hölder's axiomatization has essentially influenced the contemporary conception of the axiomatic construction of the theory of extensive magnitudes and of a formal justification of the theory of extensive measurement. A specific theory, based on a detailed analysis of the concepts of quantity, magnitude and number, can also be found in the work of Russell (1903), especially in the third volume of *Principia mathematica*. However, even this theory was not accepted as the basis for further attempts at providing the theoretico-methodological foundation of measurement. In the second period, which is represented, above all, by the work of N. R. Campbell (1920), theoretical and methodological investigations were directed exclusively from the standpoint and need of measurement in physics. Campbell was close to regarding physics as the *theory of measurement*. Although Campbell's monograph was programmatically opposed to any further extension of measurement, it nevertheless became the foundation of a general theory of measurement and the stimulation for its advancement of theoretical and methodological conceptions outside physics.

Finally, in the third period, beginning with the works of S. S. Stevens, in particular with his essay, *On the Theory of Scales of Measurement* (1946), the interest in methodological and theoretical inquiries was decisively shifted into the domain of psychologists, sociologists and economists. To them, the practice and theory of measurement becomes the central problem of the formal methodology of the empirical sciences.

The reason for this remarkable shift is undoubtedly to be found in the difficulties of a meaningful and theoretically well-founded extraphysical measurement which would play a role in the social sciences similar to the role physical measurement plays in the natural sciences. Implementation of this objective must necessarily lead to the constitution of a general theory of measurement which unifies both these kinds of measurement in a conceptual and methodological sense.[13]

The significance of this demanding project is quite obvious. Regardless of the stimuli which had incited it, every single step in this direction is not only important for the theoretical foundation of extraphysical measurement, but is also a very valuable contribution to the theory, methodology and philosophy of science.

2. MEASUREMENT

The concept of measurement, as it is understood in the most diverse connections, often with different intentions, is undoubtedly a complex concept possessing an equivocal meaning. We have already pointed this out in the Introduction. However, we have to repeat this statement to vindicate the manner that will be used in the gradual explication of the extension and content of this concept. The stipulation of a more precise definition at the very onset of our investigation would be at least premature and, at any rate, unjustified. An adequate definition of the concept of measurement, whatever its formulation, obviously has a great conceptual significance, yet it cannot be regarded as the only aim of our analysis. It is much more important for the theory of measurement if we expound what can be meaningfully said about measurement from various standpoints: if we state what are the general and specific characteristics helping us to grasp the core of the method, what its role is in the process of scientific knowledge, under what conditions measurement can be legitimately applied, and what indeed measurement is objectively. These goals can be fulfilled only in a step-by-step fashion, in a continuous confrontation with different views concerning these and other questions of the theory and practice of measurement.

2.1. THE EXPLICATION OF THE CONCEPT OF MEASUREMENT

It is quite obvious, even without a deeper acquaintance with the process of measurement – whether from our own experience with measurement procedures in experimental work, or from the study of methodological and theoretical literature – that measurement encompasses different aspects and components of an empirical and theoretical nature, which are mutually conditioned in a very complicated way. In practice, the implementation of the process of measurement appears at the foreground and contains the following: the preparation and performance of experiments employing measurements within a certain scientific area; the choice of suitable measuring operations; the construction and use of measuring instruments; and the elaboration and evaluation of the results of a measurement. On a theoretical level, the relevant problems contain, in particular, a conceptualization of the objects of measurement and of its results; a demarcation of basic concepts of the theory of measurement and of the conditions of measurability; the elucidation of the relationship between the empirical and mathematical aspects of measurement; and the constitution of a general theory of measurement. Yet not all of these aspects which, by the way, can be formulated in greater detail, are involved to the same degree: neither in each specific case of measurement nor in all theoretical analyses which can be carried out at various levels.

If we would gradually like to clarify how the concept of measurement is conceived, or what we ought to understand by it, we could investigate the various kinds of measurement or analyze the different explications of its concept.

If we were to investigate other cases of physical measurement following the exposition of measuring length and mass – for instance, the measurement of the thermal coefficient of length expansion in the theory of heat, the measurement of a temperature by means of a thermal cell in the theory of electricity, or the measurement of refraction in optics – we would very soon find out that these kinds of measurement cannot be characterized by the three analogical rules (disregarding, furthermore, the methods of measurement and the measuring instruments used). This discovery will be demonstrated in an even more drastic form, if we investigate the various procedures regarded as cases of extraphysical measurement, for example, the measurement of the intelligence quotient in psychology, utility in economics, work attitudes in sociology, student achievements in pedagogy, the meaning of words in linguistics, and the like. Then, although it is possible to attempt to formulate a doctrine by virtue of an inductive generalization – a doctrine which would express the general aspects of the concept of measurement – such an attempt is, nonetheless, futile. Such an approach has already been adopted in the past, not only by theorists and methodologists of measurements, but by scientists and engineers themselves.

It will therefore be more economical and appropriate to start our preliminary definition of the meaning of "the concept of measurement" with an analysis of explications or descriptions of this concept. It will also serve as a certain inventory of the interpretationally and terminologically controversial problems. Ordinary descriptions and explications arise from a knowledge of the nature of measurement in differentiated degrees of generality, predominantly within different scientific domains, or with regards to measurement in one discipline only. Since there are obvious differences between physical and extraphysical measurements, the richest information will be provided by an explication of the concept of measurement which is based on the confrontation of different views concerning the nature of both these fundamental, domain-oriented, types of measurement.

The following selection of conceptions, which first characterize physical and then extraphysical measurements, should not be viewed as exhibiting any special intentions and, in particular, it should not be understood as a sign of an *a priori* bias against extraphysical measurement. Even if we had offered different documentations, it might have been shown that the views concerning the nature of measurement in the social sciences are expressed in a much less determinate form then the views referring to measurement in physics.

The conception of measurement, as it is usually conceived in the context of physical measurement, can be illustrated by the following two descriptions:

(i) "The performance of quantitative descriptions (i.e., experiments by means of which we get numerical data that enable us to find out, not only the character (quality), but also the magnitude (quantity) of the observed changes), we call *measurement*. Measurement serves the aims of an objective determination of physical magnitudes . . . almost any physical measurement can be reduced to the measurement of 'lengths' (as 'concerns, in particular, all measurements where one reads off the scale the position of the pointer, filament, light index, fluid level, and the like'), for the measured magnitude is here determined in terms of 'the distance of the read position from the zero position, i.e., by a length' . . . to measure a physical magnitude means to compare it with a certain quantity of a homogeneous magnitude that has been chosen as the unit."[1]

(ii) "Measuring implies at least three distinct elements: an object, or rather a *physical system*, on which a certain operation will be performed; an observable *property* of this system whose 'value' will be determined by this operation; and an *instrument* by means of which the operation will be made . . . One can speak of a measurement only when the operation connecting these three elements results in a metric number. If no such number results, there is, at most, question of an experiment or of an observation . . ."[2]

From either of these formulations (in which the emphasis is only put on some basic aspects of measuring procedures), as well as from the conceptions that are implicitly contained in them we can draw the following characteristics:

(1) Physical measurement refers to real objects which are primarily independent of the knowing subject.

(2) From a methodological viewpoint, measurement is connected with *observation* and *experimentation*. It is connected with observation due to the fact that every measurement presupposes some observable property of the measured object. This empirical procedure is, at the same time, necessary for ascertaining numerical values with the help of the scales of measuring instruments. The relation to experimentation is conditioned by the stipulation that measurement could be conceived as a specific sort of experiment.

(3) The fundamental constituents of measurement comprise, on the one hand, observable properties (qualities) or magnitudes (quantities) of the measured objects and, on the other hand, numerically objectivized determinations of these qualities and quantities. These fundamental components can, therefore, be considered the *object* of measurement and also its *result*. The result, which expresses the magnitude of the measured object in a numerical form is at least a rational number or, if need be, a real number (if calculations are also employed). The result of a measurement consists not only of a specific numerical datum, but, as a rule, of a whole set of such data, including their statistics.

(4) A measurement procedure depends on *empirical operations* which provide the connecting link between both basic elements, on the employment of *measuring instruments* which facilitate the specification of numerical values of the measured properties by reading the given data off the scales of these measuring devices, as well as on the existence of a suitable *unit of measurement* which is homogeneous with the object of measurement. The very process of measurement rests with *comparing* the measured property with the magnitude of the unit of measurement. If we use a measuring instrument, then every physical measurement can be practically reduced to a length measurement.[3] These conditions of operationalization can be summarized with the help of certain *rules* that are specific for certain sorts of measurement.

For an extraphysical measurement we shall likewise give two explications which are, however, interpretationally different, in spite of the apparent terminological agreement:

(i) "Measurement has two components: (a) *Measuring instruments*: ordinarily these are prepared scales, arranged into a transparent and intelligible form which we put before questioned persons or use in grading and point assignments ... (b) *Measurement proper*: whether this is carried out by the questioned person or by the observer, there is still a whole range of special mental aspects, such as measurement capability (the objectivity of the observer or the ability to estimate the sign level by the questioned person) and a comprehension of the instructions for measurement, etc."[4]

(ii) "In the broadest sense of the word, measurement might be regarded as a classification of objects and events in which a certain sign (numeral, letter, word, etc.) is assigned to each defined class."[5]

From these two descriptions we can now derive – in comparison with the above-mentioned characteristics of physical measurement – further specifications. Some of these specifications are in agreement with each other, while others differ.

(1) In contrast to physical measurement, extraphysical measurement refers, conceptually and operationally, to man; or more precisely, to such of his subjective properties (feelings, attitudes, desires etc.) which basically cannot be physically measured. It does not, however, refer to those properties of man (height, weight etc.,) which are not specific for a measurement in the social sciences. This difference is an objective expression of the distinction between the investigated subject matter of physics and that of various social scientific areas.

(2) Extraphysical measurement is methodologically tied up with *classification* and hence, it is assumed, with observation too. It is even held to be a *kind of classification*[6] and this characteristic is then, generally, also applied to a measurement referring to arbitrary objects and phenomena, and therefore to physical measurement as well. The position, according to which measurement has closer relations to classification rather than to observation and experiment, already implies diverse consequences of a methodological nature. In the first case, the emphasis is put on the conceptual aspect of measurement, in the second, on the empirical level and on an operational characterization of measurement.

(3) Extraphysical measurement is likewise based on the assumption of two basic elements – the object and result of measurement, though this result is already identified with *numeral symbolization*.[7] The difference between numerical and numeral expression is, of course, of a quite essential nature; it is connected with diverse views on the extension of the concept of measurement.

(4) Operationally, the measurement procedure is only very ambiguously determined, and although it is also characterized with the help of measuring instruments and scales, these instruments are here understood to be something essentially different. Not only are scales acknowledged as measuring instruments, but so also are "observation, questionnaire, and interview."[8] However, the scales which are so often mentioned, particularly in the conception of so-called behavioral and social measurement, are not the scales of measuring

instruments, but conceptual means representing the results of a certain type or level of measurement – the so-called *scales of measurement*. As a consequence of this distinctive difference, from an operational standpoint, the very process of an extraphysical measurement proceeds in a different manner, namely on the basis of subjective factors.

Without evaluating thus far the appropriateness of all the single conceptions which have been stated in these explications, it is quite evident that, in spite of considerable divergencies, they agree in characterizing the process of measurement by means of three basic components: the object of measurement, the results of measurement, and certain mediating empirical operations. Yet, are these components sufficient and, at the same time, relevant for the realization and justification of the method of measurement in all its complexity and in all spheres of its application, irrespective of any concrete interpretation of these components? So far, without a further detailed analysis, we cannot give a satisfactory answer to this principal question – an answer which would refer to all sorts of measurement.

In this connection, let us now consider only the role of empirical operations. From the standpoint of operationalism, this component will be precisely regarded as the essence of measurement. This viewpoint neglects the conceptual sides of measurement and overestimates its empirical aspects, in particular the way of operationalizing measurement procedures. Thus, as to its consequences, it is also in conflict with historically-tested and theoretically-justified kinds of measurement. Let us only consider – so far without a detailed analysis and justification (see Section 6.2)[9] – the difference between so-called *fundamental* (basic) and *derived* measurement.

Fundamental measurement of some magnitude is usually understood as that measurement which does not contain any previous measurement of another magnitude. Therefore, magnitudes such as length, mass or time, for example, are measured in a fundamental way. According to this, a derived measurement is that measurement of some magnitude which depends on the fundamental measurements of other magnitudes. A common example of a derived measurement is the measurement of density. In this case we establish numerical values of the density of materials or (regular) bodies by *calculation*, on the basis of their volume and mass, without having to perform any empirical operations. Is, then, the measurement of density an empirical procedure in the correct sense of the term? A similar question can be set forth in other cases as well. For instance, if we measure the volume of regular bodies. Numerical values of this magnitude can also be acquired without the use of specific empirical operations, by means of calculation based on the measured length values. Can we, therefore, from the operationalist viewpoint, regard derived measurements as measurements? Some authors,[10] who consistently think this viewpoint to the end, arrive at an evidently doubtful conclusion that this is not the case. This means, however, that density or volume are physical magnitudes which are not measurable almost *ex definitione*; they are in fact determined by calculation. Such a point of view, which *narrows down* the extension of the concept of measurement exclusively to the fundamental measurement, is of course hardly compatible with the historical practice of physical measurement of these and many other derived measurable magnitudes.

Equally one-sided is the position according to which empirical operations are not an essential ingredient of a measurement. This position is, on the contrary, based on the absolutization of mathematical characteristics of measurement. Also this position, which disproportionately *extends* the scope of measurement, leads to unacceptable consequences, namely that every process of numerical expression could be taken for a measurement: not only the determination of numerosity (numerousness, cardinal number) of a certain set of objects by counting its elements – a position frequently adopted[11] – but also the diversified forms of calculations – a position which is inconsistently rejected.

Measuring operations play a role that is actually much more complicated and complex. They cannot be judged exclusively from either side – exclusively *purely empirically* or *purely mathematically*. Their irreducible function in the process of measurement consists of mutual connections between their conceptual and operational aspects, in semantically meaningful and empirically realizable correspondence with numerical operations, as well as of a suitable specification which depends on the nature of the object of measurement, measuring instruments used, the ways of measurement, and the like.

2.2. THE DEFINITION OF THE CONCEPT OF MEASUREMENT

The relevance of the other two components for the characterization of the method of measurement (although, even here one finds antagonistic views concerning the question of what is specifically the object of measurement and at what level the results of measurement are numerically expressed) is no more a matter of controversy, as can be attested by the various definitions of this concept.

The majority of definitions agree in subordinating the concept of measurement under the concept of a procedure (process or method), or more precisely, under the concept of the procedure of an assignment (or representation). Furthermore, they agree in giving different specific differences for this superordinated concept. Depending on both of these basic elements – on the object and the result of measurement – these specifications are related first of all to the nature, aim and conditions under which this procedure can be materialized. They determine the procedure as an assignment of *numerical* expressions to the measured objects, thus serving their *representation* on the basis of *laws* or *rules*.

In a systematic fashion, the first time we encounter this manner of defining the concept of measurement (which need not always contain all three basic specifications) is in the works of N. R. Campbell. His views – in both a positive and negative sense – have essentially influenced all further attempts to grasp the principal characteristics of the concept of measurement by a simple classical definition. Campbell defines the concept of measurement in various ways:

(i) as "the assignment of numerals to represent properties",
(ii) as "the process of assigning numbers to represent qualities",
(iii) as "the assignment of numerals to represent properties according to scientific laws",
(iv) as "the assignment of numerals to things so as to represent facts or conventions about them".[12]

All these definitions evidently agree in explicitly mentioning the representational function of the numerical or numeral assignment. By the justification of this function of numerical representation in the process of measurement, Campbell has established the foundations of the so-called *representational* theory of measurement. According to this theory which, to date, has been accepted almost unanimously, the numerical result of a measurement enables us – only if certain conditions are met, of course – to derive meaningfully relevant conclusions concerning certain properties of the object of measurement.

Yet, aside from this undoubtedly very positive contribution, we must also point out the considerably negative consequences of Campbell's conception, which are due to the fact that in the above-mentioned definitions, the terms *number* and *numeral* are used synonymously with different meanings. In addition, a new term, *Number*, is introduced, which differs in content from the previous two terms.

What is the meaning which Campbell's[13] interpretation gives to these three terms? The term *Number* is regarded by him as a mathematical concept in the sense of Russell's definition, i.e., as a class of classes which have the same numerousness and the same cardinal number. Under the term *number*, he understands a certain physical magnitude, more precisely, a physical magnitude fundamentally measurable by counting. This specific magnitude – the so-called 'physical number' – differs from all other fundamentally measurable magnitudes in three ways. In contrast with any other basic magnitude, for instance, with time, it has no dimension. However, since every number is always a number of something, for example, a number of crystals or a number of cleaning ladies (in this case, obviously, with the meaning of an *amount*), two different numbers – as in the case, for instance, of length and time – are two different magnitudes. In the second place, therefore, every number changes, with the change of the unit (evidently, in the sense of a singular object of a definite kind, not in the sense of a measurement unit), into another magnitude. Finally, as the third point, only for 'physical numbers' does multiplication have a physically meaningful interpretation. The term *numeral* is then interpreted by Campbell as a material or a quasi-material symbol, as an arbitrary symbol, as a name of a *number*, and sometimes even as a name for *Number*. At the same time Campbell admits that numerals can be ordered into a standard series and mutually counted, so that one can also perform all basic arithmetical operations with them, irrespective of their relationship to Numbers.

In Campbell's conception, the differentiation between these three terms with diverse semantical interpretation is conditioned by a purely empirical and strictly nominalistic position which rests ontologically on the acceptance of the reality of numbers as physical magnitudes of a specific type. The mathematical concept of number, namely *Number*, concerning which "the plain man knows nothing . . . nor indeed did the most skilled mathematician know anything about them until a period within living memory,"[14] is rejected, for its definition is tied in with Russell's logicism which Campbell repudiates for philosophical reasons. Therefore, he arrives at the view that, in the theory and practice of measurement, it is sufficient to deal exclusively with the notions of *number* and *numeral* (and this, even in the case of numerical calculations that are necessary for ascertaining the values of derivatively measured magnitudes) of course if we introduce fractional numerals or irrational numerals as well.

However, with respect to these numerals, Campbell inconsistently asserts that they represent *irrational Numbers.*[15] There are, of course, other inconsistencies committed by him in his exposition. Since we are not going to give a detailed analysis and evaluation of Campbell's philosophy of mathematics, the unjustifiable character of which is perhaps beyond any doubt, we do not consider it necessary to point out further conflicts in his doctrine. Only two questions are important to us in this connection: How can one explicate the terms *number* and *numeral* in Campbell's definitions of the concept of measurement? In what way have these definitions influenced the conception of measurement in later interpretations?

First of all, it is obvious that both these terms are tied together by a very unique philosophy of mathematics and that, unless we accept it, we cannot assign the same meaning to them. When Campbell talks about numbers and numerals, he does not understand these concepts exactly as they are commonly conceived. However, if we consider the fact that in the process of measurement, Campbell distinguishes two different levels (empirical and conceptual), that he regards numbers as 'physical objects' and numerals as their linguistic expression, and, in addition, that he accepts the possibility of performing with numerals all the basic arithmetical operations, then we must admit that he indeed treats them, disregarding what he says about them, exactly as if they were numbers in the common sense of the term.

Yet, if one overlooks or fails to notice the specificity of Campbell's conception and if one transfers his interpretation of the terms *number* and *numeral* beyond the frame of the conceptual network of his theory, there necessarily arises a meaning contamination. The failure to keep the difference between *numbers* and *numerals*, or a confusion between them leads to questionable conclusions – this is what particularly weighs down Stevens' position.

S. S. Stevens receives Campbell's definition of the concept of measurement and modifies it not only by omitting an explicit expression of the representational function but also by adding his formulation of the conditions under which measurement is realizable. On Stevens' view, one can hold measurement to be every assignment of numerals to objects or events, or, if you like, to aspects of objects or events, according to rules – according to any rule.[16]

This terminological agreement, which, however, does not correspond to an appropriate conceptual agreement, is not helpful for grasping the conceptual linkage between these two authors, by means of which Stevens tries to support his own position. As it appears, Stevens is even convinced that he makes more precise Campbell's "very liberal" definition of the concept of measurement as "the assignment of numerals to things so as to represent facts and conventions about them,"[17] by putting the emphasis in the definiens on the third specific difference of the concept of assignment. He assumes that the problem, what indeed measurement is and what it is not, may be reduced to a simple question: What are the rules under which numerals are assigned? This, however, is far from explaining the essential properties of the concept of measurement, or the conditions under which measurement is meaningfully realizable in practice. If numerals can also represent conventions referring to things, the rules of assignment might be very arbitrary. Hence, Stevens' specification of the concept of measurement need not lead to its greater precision. If, on the basis of Campbell's definition, one admits the possibility of a purely conventional formulation of the rule of assignment, the extension of the concept of measurement can be, on the contrary, increased in an arbitrary manner. Then anything can be conceived as measurement, for conventionally one can always find or stipulate some rule, with the help of which numerals could be assigned to diversified aspects of things. This, however, will have very little in common with measurement in the correct sense of the term. Although the possibility of such a subjective comprehension of the rules of assignment complies with Stevens' striving for the employment of a measurement procedure to the greatest possible extent, it is definitely in conflict – in spite of the terminological agreement – with the strictly objective, physically oriented conception of measurement held by Campbell.

In Stevens' definition, the conception of the assignment of numerals also serves the same purpose. It should add to the procedure of naming by means of numerals (numeral designation) "the dignity implied by the term measurement."[18] The assignment of numerals to aspects of the measured objects can only lead, however, to numeral results. Then, of course, numerals are not used as proper names, the denotata of which are either physical magnitudes or mathematical entities, but only as certain symbols which can be interpreted in a most diverse way. Since Stevens does not accept Campbell's philosophy of mathematics, he should not appeal to his definition of the concept of measurement. When both talk about numbers and numerals, each of them understands something different, Stevens certainly conceives the term *number*, not as a physical magnitude, but as a mathematical concept, i.e., in the sense of Campbell's term *Number*.

The mixture of the terms *numeral* and *number* in the definitions of the concept of measurement, conditioned by an unacceptable semantic confusion between the symbol and its denotatum, designation and designatum, thus leads, in Stevens' position, to entirely different consequences than from Campbell's viewpoint. Campbell always regards *numerals* only as names of numbers – quantitatively determined physical magnitudes – so that in the final account we can always interpret them in the same sense in which we ordinarily conceive numbers as mathematical entities. Stevens, however, understands *numerals* in two completely different meanings: as names of mathematical entities and as entirely arbitrary symbols.

It is characteristic of other definitions as well, in conformity with the representational theory of measurement, that they emphasize the relation between the object of measurement and its result, just as mutual connections between them. These definitions do not mention – at least when they refer to the concept of measurement in a narrow sense – the assignment of numerals. Only a numerical assignment is taken into account. Instead of only an indefinitely formulated modification of Campbell's definition "according to rules", these definitions also specify the manner of numerical assignment by means of numerical relations and operations referring to empirical relations and operations. This way of defining the concept of measurement, which can be illustrated by the following definition:

> measurement consists in assigning numbers to things in such a way that certain operations on and relations among the assigned numbers come to 'correspond to' or 'represent' observable relations and operations on the things to which they are assigned,[19]

is influenced by the formal conception of the theory of measurement, represented, in particular, by the works of Scott and Suppes (1958), Suppes and Zinnes (1963), Krantz *et al.* (1971) and Kanger (1972). Based on the results of the mathematical theory of models, as formulated by Tarski, measurement is then understood as a *homomorphic mapping of a certain empirical relational system* (empirical structure) *onto some numerical relational system* (numerical structure).

The discussion of this current, widespread position, which has been developed in connection with the study of the formal foundations of the theory of measurement, will enable us to pass judgment on the nature of the object of measurement, as well as on the function of empirical operations.

In general, we render the concept of a relational system and relational structure $\mathscr{S}$, in accord with the theory of systems,[20] as some ordered n-tuple of sets

$$\mathscr{S} = \langle D, R \rangle,$$

where D is a non-empty (finite or infinite) set of objects of an arbitrary kind, and R is a non-empty (finite or infinite) set of n-placed relations ($n = 1, 2, 3, \ldots$) or operations; this, of course, contains properties too (for $n = 1$).

The difference between the concepts of a relational system and a relational structure, which are also sometimes identified, reflects only the degree of abstraction with which the given problems are judged. The concept of a relational system is more concrete: in this case both sets $\boldsymbol{D}$ and $\boldsymbol{R}$ are explicitly given. The concept of a relational structure is more abstract: it is explicitly defined only through the set $\boldsymbol{R}$. This set can be generally characterized as an alternative of two subsets

$$\boldsymbol{R} = \{R \vee O\},$$

where $\boldsymbol{R}$ is a set of relations $R_1^n, R_2^n, \ldots, R_k^n$, and $\boldsymbol{O}$ is a set of operations $O_1^n, O_2^n, \ldots, O_j^n$ $(n = 1, 2, 3, \ldots)$. If the set $\boldsymbol{R}$ contains only the subset $\boldsymbol{R}$, we consider the relational system $\mathscr{S}$ as *natural*; if it contains the subset $\boldsymbol{O}$, the relational system is *artificial*. This distinction rests on a specific comprehension of operations as a result of human activity and as a theoretico-practical reflexion of definite objective bonds. Therefore, it primarily characterizes only artificial relational systems.

According to the way the set $\boldsymbol{D}$ is given – which determines the nature of the set $\boldsymbol{R}$ as well – we distinguish then, in principle, two sorts of relational systems (structures): numerical and empirical.

A numerical relational system

$$\mathscr{N} = \langle N, \boldsymbol{R}_N \rangle,$$

which is an artificial system, contains the set of numerical objects N and the set $\boldsymbol{R}_N$, the elements of which are numerical relations $R_{N_1}^n, R_{N_2}^n, \ldots, R_{N_k}^n$, and numerical operations $O_{N_1}^n, O_{N_2}^n, \ldots, O_{N_j}^n$ $(n = 1, 2, 3, \ldots)$.

An empirical relational system

$$\mathscr{E} = \langle E, \boldsymbol{R}_E \rangle,$$

which is primarily a natural system, contains the set of empirical objects E and the set of empirical relations $\boldsymbol{R}_E$. However, it does not objectively contain any subset of operations.

On the basis of this conception, the concept of measurement $\mathscr{M}$ can be characterized as a binary relation

$$\mathscr{M}(\mathscr{E}, \mathscr{N}),$$

that holds between some empirical relational system and some numerical relational system; or perhaps as an order triad

$$\langle \mathscr{E}, \mathscr{N}, \Phi \rangle,$$

where Φ is a definite homomorphic mapping of $\mathscr{E}$ onto $\mathscr{N}$.

But even this general definition of the concept of measurement from the structural standpoint, is not entirely free of problems, for the assumption of a homomorphic mapping between both relational systems is, strictly speaking, in conflict with the diverse characters of the sets $\boldsymbol{R}_E$ and $\boldsymbol{R}_N$. The elements of the set of numerical relations $\boldsymbol{R}_N$ are comprised of relations between numbers, on one hand, and the operations between them, on the other hand. Yet, in empirical relational systems, there cannot primarily occur any operations besides empirical relations.

What are the conditions, therefore, under which we can accept the existence of a certain, indeed symmetrical, homomorphism also based on the correspondence between numerical and empirical operations? Evidently, we can do so only if we admit that empirical operations can be introduced into the system $\mathscr{E}$, at least secondarily. This possibility is necessary for the achievement of the required relationship between both relational systems. The point of departure for every measurement is the knowledge of objectively existing relations between the objects and phenomena of objective reality. On the basis of this knowledge we then start to look for a certain numerical expression. At this initial stage, the mapping of an empirical relational system onto a numerical relational system is a homomorphism based only on the correspondence between empirical and numerical objects, between empirical and numerical relations. However, since the numerical relational system is also defined by operations with numbers, in the second stage we proceed in the reverse order. We strive to find suitable empirical counterparts to these numerical operations – counterparts susceptible of meaningful interpretations. If we succeed in finding empirical operations that have properties structurally analogical with the properties of numerical operations, then we can talk about a homomorphic mapping of the system $\mathscr{E}$ into the system $\mathscr{N}$ in the strict sense. Only under this assumption can we derive certified conclusions concerning the mutual relations between both relational systems and judge from the properties of elements of the numerical set N to the analogical properties of elements of the empirical set $\boldsymbol{E}$. Of course, if we do not succeed, for whatever reason, we must attempt to provide some other solution that would, nevertheless, be sufficiently justified theoretically.

At the same time, we have shown through this explanation that the object of measurement, formally represented by the concept of an empirical relational system, is always somewhat conceptualized. This observation also serves as a serious argument against the underestimation of the role of theory in the practice of measurement.

2.3. THE SUBJECT MATTER, FUNCTION AND SCOPE OF MEASUREMENT

What conclusions concerning the nature of measurement can we now deduce from the previous exposition?

The difficulties which we have encountered in defining the nature of an empirical relational system lead us, first of all, to posing the question of the *subject matter of measurement*. If we speak, for instance, about a length measurement and express the result by the statement, "This board is 5 m long," or, "The length of this board is 5 m," the numerical expression refers to the length, i.e., to a definite property of the measured object. If we ascertain the size of a certain population, say, the number of persons who had reached the age of 50, and if we express the result in the statement, "There are 10 persons in the population in question," the numerical expression refers again to the size of this population and not to the age of individual persons. Obviously, there are some differences between these two cases. In one we are considering the property of an object and in the other, the properties of a class of objects. At the same time, what is the role played by these same objects? If we measure length, we can obviously measure it only on a definite object, or between two objects, and the like. If we ascertain the number of persons of the same age, we must know these persons in some way. How then is the set E defined? Are its elements some empirical objects, empirical collections of objects, or the properties or relations between them? This fundamental question may be answered in two ways, according to our consideration of the conceptual and operational aspects of measurement. Conceptually, we can assume such a division of the universe that the elements of the empirical set E would be precisely those objects to which the elements of the numerical set N are immediately assigned. As it appears so far, the set E would thus contain only properties of things and phenomena or, eventually, the relations between them, but not these things and phenomena as such. Such a division of the universe, and consequently the definition of the set E too, is, of course, unthinkable from the operational standpoint. Every measurement is always performed on objectively existing things and phenomena. This is also evident for philosophical reasons: in objective reality there exist no properties or relations as such, but always only spatiotemporally determined things and phenomena having certain properties – things and phenomena which occur in definite relations independently of our conceptual selection of the properties and relations between them.

Often, there exist very complex connections between the measured objects, *on* which or *with the help of* which we measure something (constituting the objective foundation of every measurement), and the features which we measure or *with the help of which* we measure these objects. The delimitation of elements of the set E must, in any case, take account of the mutual connections between the conceptual and operational aspects of measurement and their theoretical and philosophical foundations. It is evident, even without a more detailed analysis of these complex problems, that that *which* we measure, and that *on what* or *by means of* which we measure, are not to be identified, nor viewed as mutually isolated factors.

In any case, our distinction between the *measured object* and the *object of measurement* may now be taken as being already justified.

Secondly, we might ponder the question of what place measurement has in the system of scientific procedures. As we have already mentioned (Section 2.1), we cannot find an unambiguous answer to this question. In principle, there exist three different positions with regard to the various aspects of measurement, its function in the process of knowledge, and its practical use.

There is a position held in physics (mainly under the influence of its experimental practice), according to which measurement, together with observation and experiment, is an *empirical* method or, even more narrowly, only an operational procedure. This position is also held by the neopositivists in their classification of methods into empirical and theoretical. Among these can be found, on one hand, the methods of deduction and definitional construction of scientific theories, such as, for instance, classification, definition, axiomatization, etc., and on the other hand, prediction, explanation, and verification.

Another position is held by many representatives of extraphysical measurement. They assume, to the contrary, that measurement is only a refined form of classification and, hence, is essentially a *theoretical*, or perhaps a *conceptual* method.

However, if in the process of scientific knowledge we take into account the mediating function of measurement between theory and practice, between empirical knowledge and its mathematical expression, then measurement can be rendered as an *empirico-mathematical* method.

According to our opinion, it is precisely this third conception that is the most truthful. It is, of course, possible in principle to consider measurement as an empirical method – naturally, if we do not neglect its theoretical presuppositions – and this viewpoint can often be encountered. Yet it is incorrect to assume that measurement is only a conceptual method.

If measurement is understood as an empirical method, this only reflects the manner in which the measurement of values of different magnitudes, or of one and the same magnitude, is carried out in practice, with the help of various measuring operations and with dependence upon the measuring instruments employed, as well as on the required values of measured magnitudes and other conditions that are connected with the technical realization of experiments employing measurement. Then the method of measurement is held to be, not only every general measurement procedure which is applied in measuring various magnitudes (for example, a direct, a substitutional, or a compensational procedure) but also every specific procedure used in measuring one and the same magnitude (for instance, deflection or zero procedures in measuring resistance), and even very specialized measurement techniques, conditioned by the choice of measuring instruments (for example, the pycnometer, the areometer, the immersion body, or the Mohr-Westphal balance in measuring the density of fluids). Evidently, this conception overlooks the essential difference between the method and the ways of its operationalization. It does not adequately grasp either the methodological function of measurement in relation to other methods or the various levels, sorts and forms of measurement.

The reduction of problems of measurement to only their empirical aspects – and this is another philosophically very significant consequence of this conception of which the experimenter need not be immediately aware – is, of course, acceptable only from the standpoint of pure empiricism, according to which there exist exclusively either empirical or theoretical methods. For factual reasons, this strictly dichotomic classification of methods, which ignores mutual bonds and relations between different methods, is untenable precisely in the case of the method of measurement. Yet, how can one leave undisturbed the original classification and at the same time cope with the mathematical method of measurement, with its conceptual presuppositions? Obviously, only a dualistic way out of this dilemma can be found: either, as before, one regards measurement as an empirical method; but isolates its *non-empirical* components and turns them into a specific conceptual method; or one starts from a definite theoretical method and then renders measurement its empirical correlate.

In the first case, the method of measurement is accompanied by the method of *metrization*,[21] which is constituted on the conceptual plane and which defines the conditions – the so-called *metrical* conditions –

under which one can move from classificational concepts to metrical ones (6.3). The introduction of the notion of metrization undoubtedly represents a significant and positive departure from the position of operationalism, as well as from a merely phenomenological viewpoint which even emphasizes that the most basic sort of measurement consists in the sheer reading of data from the scales of measuring instruments. Of course, if both of these methods are put into a mutual correlation without jointly representing a unified concept of measurement in both of its basic aspects, if measurement is regarded only as an empirico-mathematical method, then there always exists here a real possibility of treating them in isolation, irrespective of their mutual bonds. This often happens in reality. Then the discussion revolves around questions such as whether the metrization of a definite concept is a necessary prerequisite of its measurability, or whether something can be measured without meeting the conditions of metrization. According to our opinion, these questions are put forward in an improper way.

In the second case, the opposite tendency comes to the fore: the original empirical starting point retreats into the background as to the mathematical results. Measurement is only an empirical counterpart of *quantification* or a definition of measure.[22] The priority of quantification is then justified by the claim that in mathematics there exist quantitative theories which do not refer to the empirical operation called *measurement*. Although this argument as such is not convincing, it is sufficient ground for finding out that it rests on a lack of conceptual clarity with respect to quantification in the empirical sciences and in mathematics. The consequences, which follow from taking the concept of measurement as lacking unity, now manifest themselves in the absolutization of its mathematical aspects. We encounter this phenomenon especially in purely structurally oriented analyses, which are almost exclusively concentrated on the axiomatization of various levels or sorts of measurement, disregarding questions such as, whether these kinds of measurement are meaningfully realizable in practice, or whether they have some corresponding empirical correlate to which a character of measurement could be attributed.

The assertion that measurement is a sort of classification, or, rather on the contrary, that classification is already measurement, follows directly from Stevens' definition of the concept of measurement: namely, that classification meets the condition of "the assignment of numerals according to some rule," for example, in the enumeration of items in a catalogue, and so it is distinguished from merely accidental assignment where no rule is employed. This justification might be plausible only if we accept Stevens' definition of the concept of measurement. Stevens himself is aware of its problematic nature, particularly in the case of its applicability to classification. Therefore, he defends himself against a possible criticism by turning the entire discussion onto a purely pragmatical level and claims that "the question whether the process of classification . . . is measurement, is one of those semantical problems which depend on our taste."[23]

Obviously, it is not so important whether in classification we use the assignment of numerals or not, whether, for example, we classify books according to the Dewey Decimal Classification, or according to some other procedure. What is decisive is only the fact that classification is primarily a theoretical, and not an empirical or empirico-mathematical method. From the logical standpoint, classification can be characterized as an analysis of a certain class of objects into various, mutually disjoint, subclasses, the union of which is equal to the initial classificational collection (in the case of an analytical classification). We might also conceive of it as the reverse procedure which meets the same conditions, in the case of a synthetic classification. Classification is immediately connected only with the constitution of classificational (qualitative) concepts, not, however, with metrical (quantitative) concepts. It is a presupposition, though not an instance, of measurement only in this intervening sense. However, in the same way, it is also a presupposition of many other methods. If we regarded every presupposition of a measurement as the measurement itself, we would have to say the same, for instance, about an identification. The claim that classification is an instance of measurement, or that measurement is a sort of classification, is therefore wrong, for it implies an unwarranted and unjustified broadening of the extension of the concept of measurement.

On the basis of previous discussions, we can now derive the following conclusions and terminological conventions:

From the methodological standpoint, we regard measurement as an *empirico-mathematical method* that uniformly comprises two components, one, empirico-theoretical the other, conceptually-mathematical; in either case, theoretically justified with different degrees of generality.

If we emphasize only the conceptually-mathematical component, we shall talk about *metrization*. If, to the contrary, we consider its empirico-operational aspect, above all, concretely performed measurements, we shall talk about *measuring procedures* (procedures of measurement) on a more general level, or, more specifically, about *methods of measurement*.

Thirdly, we can pass judgment on the question of whether and under what conditions we can talk, in the definition of the concept of measurement, about the *assignment of numerals*.

If we are to analyze these problems, tied up with the delimitation of the object of measurement (see Section 3.3), we have to distinguish three cases: either the term *numeral* is comprehended as an entirely arbitrary symbol; or it is held to be a name of a number which, of course, does not yet express any quantitative determination in the strict sense of the term; or we mean a name of a number that already expresses such determination.

From these three possible interpretations, different consequences follow for the determination of the extension of the concept of measurement.

If, in the definition of the concept of measurement, we admit the term *the assignment of numerals* in the first sense, the extension of this concept will be almost unlimited, irrespective of the restricting condition "according to rules". If we assume that numerals designate numbers which, however, do not yet express any quantitative determination of the measured objects in the strict sense of the term, then the extension of the concept of measurement will already be limited by the conditions that, in analogy with metrical conditions, we designate as *topological*. Speaking in anticipation, these conditions of *topologization* (Section 6.3) characterize necessary presuppositions on the basis of which one can proceed from classificational (qualitative) concepts to topological (comparative) concepts.

If numerals do designate numbers which already express quantitative determinations in the strict sense of the term, the extension of the concept of measurement will be even more restricted, being contingent upon the conditions of metrization.

In what extension should we then conceive of the concept of measurement? Under the influence of the extensions of measurements in the social sciences, two different levels of measurement are usually differentiated: the so-called *lower* level (classification, numbering, numeral designation and ordering) and the so-called *higher* level, which corresponds to the scope of measurement as it is understood in the context of physical measurement. These different levels are also terminologically fixed by virtue of the differentiation between qualitative and quantitative measurement, or between classical and liberalized theories of measurement.[24] In the same sense, we shall speak for the time being about the *wider* and *narrower* comprehensions of measurement.

The delimitation of the extension of the concept of measurement in the narrower sense of the word, as far as we treat it only within this classification, is generally comprehended with full agreement: it must meet the conditions of metrization and interpretation of the term *the assignment of numerals* in its third meaning. When establishing the extension of the concept of measurement in the wider sense of the word, we encounter two positions.

According to the first, it is required that topological conditions be met and the term *the assignment of numerals* be interpreted in its second meaning.[25] According to the second, even an arbitrary assignment of numerals is admissible, so far as it meets the condition "according to rules".[26]

We have already partly outlined our position with regard to these problems. Since, from a structural standpoint, we have defined the concept of measurement as a homomorphic mapping of some empirical relational system onto a definite *numerical* relational system, it is obvious that we consider the use of the term *the assignment of numerals* as a correct one only if it is a name of a numerical mapping. Then it is also unnecessary to distinguish in the definitions of the concept of measurement between its *numeral* and *numerical* specification. Of course, even if we talk only about a numerical mapping, it is not yet clear whether we have in mind one or other or both of the further interpretations of the term *numeral*.

In addition, the assignment of numerals is, on our view, entirely arbitrary, regardless of whether it is restricted by some consistent and systematically applicable rule. We might, of course, easily imagine some, even very complicated rule. For example, a rule with the help of which one can assign numerals to the paving stones on the Old Town Square with regard to the number of persons who had stepped on them with their left foot in the course of one year. Yet, this consistent and systematically applicable rule of assignment of numerals would hardly be regarded as meaningful. Even less could it be regarded as a sufficient precondition of measurability on this, already *lower* level. The elimination of numeral assignment – and consequently also of classification, numbering, numerical designation – from the extension of the concept of measurement on the *lower* level, is further implied by the fact that, from the methodological standpoint, we have delimited the concept of measurement as an empirico-mathematical method. The extension of the concept of measurement on this level is thus restricted only to the cases of numerical mapping which satisfy the conditions of topologization.

So far, we cannot give any satisfactory answer to the question of whether one might also talk about measurement on such a restricted *lower* level, or whether one should understand under the concept of measurement only a measurement on a *higher* level. However, if we accept as legitimate the position of the *narrower* conception of measurement, and if, in addition, we succeed in justifying this position by a conclusive argument, such a decision cannot be regarded as a mere terminological convention.

3. MAGNITUDES

The distinction between the measured object and the object of measurement is only the first step toward an appropriate specification of what is actually the subject matter of measurement. This problem can be investigated from various standpoints. We might consider the necessary ontologico-gnosiological presuppositions, the objective conditions under which we can say about some property of the measured object that it is an object of measurement and so it admits the possibility of an empirically relevant numerical mapping. These problems must inevitably lead us towards the differentiation of qualitative and quantitative aspects and towards the distinction between quantity and quality. The object of measurement must also be judged from the standpoint of the mutual relationship between the empirical and numerical aspects of measurement. It can also be analyzed from the viewpoint of differentiating concepts as classificational, topological, and metrical. All these questions, whether we judge them from one or several points of view, lead us towards the basic objective: to conceptualize the object of measurement so that it might satisfy the theory and practice of measurement.

This problem is not an easy one. Its demarcation depends, above all, on the extension of the concept of measurement: on the conception of measurement in the wider, narrower, or even the widest sense. It too depends on different forms and kinds of physical and extraphysical measurement. We must consider this problem at various levels of abstraction. In addition, we must not neglect diverse theoretical and philosophical viewpoints. If we take into account all these circumstances, the enumeration of which could be extended by still further conditions and connections, we might quite naturally set forth the question of whether the object of measurement can be conceptually grasped at all and terminologically fixed so that it would meet these different aspects, connections, and very diverse views concerning the nature of measurement.

What then is the object of measurement? Is it comprised of any properties, in which case we admit a very wide conception of measurement, or is it comprised only of quantitative properties, in which case we accept the narrower point of view? Is this classification sufficient? Can there not exist further possibilities? Can we isolate the object of measurement from the result of measurement? On what level is this possible, and/or impossible? Could quantities and magnitudes be regarded as objects of measurement? Are we to understand these concepts as having the same meaning, or should we differentiate them conceptually? What, indeed, is a magnitude? Can we accept the position of physics and transfer the concept of physical magnitude into extraphysical domains without further modifications? Are all measurable properties in physics magnitudes?

We shall now try to elucidate these questions on a level which is, on the basis of the foregoing exposition, possible and also necessary for the preliminary justification of our standpoint. For factual reasons and, to some extent, for terminological reasons too, we accept as the most convenient assumption that, on the empirico-mathematical level of measurement, the object of measurement is comprised of various magnitudes.

When analyzing these problems, we shall be compelled to take into account the relationship between this concept and the related concepts of quantity, quality, property, and number, and to reconcile ourselves with the different interpretations of these notions, especially with the concept of magnitude in mathematics. Just as important will be the investigation of the conditions under which one might utilize the information about physical magnitudes in order to constitute a general concept of magnitude that would be acceptable – with necessary specifications, of course – for the theory of extraphysical measurement as well. Since the concept of physical magnitude has a precise and clearly defined meaning, we may expect that this terminological extension will encounter various objections. Of course, it would be possible to confine the concept of magnitude only to objects of physical measurement, and to introduce some other term for objects of extraphysical measurement, either some of those in common use (e.g., sign, indicator, variable, continuum, index, dimension, and the like) or some new term that is not too loaded in its meaning. Yet, even this alternative is problematic; obviously, because of the patent ambiguity and obscurity of these terms. To introduce an entirely new term would mean only to multiply this variability. The decisive motive for the proposed terminological convention lies in our desire to express the unity – unity in diversity of course – of the general theory of measurement with the common theoretical, methodological, and philosophical foundations.

3.1. QUANTITIES, MAGNITUDES, NUMBERS: A HISTORICAL EXCURSION

If we want to explain the concepts of quantity and magnitude, in particular in relation to the concept of number, we may be helped, above all, by making a concise historical excursion into the history of mathematics in terms of which we should like to justify the difference between the conceptions of magnitude in mathematics and in physics.

For Greek mathematics there existed at first a very essential difference between the notions of number (*arithmos*) and magnitude (*megethos*), based on the differentiation of arithmetic and geometry. This difference was also reflected philosophically – in the arithmetical comprehension of the world by the Pythagoreans and in the geometrical conception of the Platonic school. The concept of number was confined only to whole numbers composed of units (*monas*) and to rational fractions which were understood as operations with these integers. The concept of magnitude comprised various geometrical formations such as lines, positions of a body, and the like. These basic species of quantitative concepts are also contained in Aristotle's categorial system as specific concepts of the category of quantity (*poson*), defined as "what is divisible into two or more constituent parts, each being naturally a unit or a 'this-something'."[1] The category of quantity reflects either an amount, in which case the quantitative properties can be counted, or the size (magnitude), in which case they can be measured. Counting refers to numbers which are held to be discrete entities, for their parts are not contiguous and do not occupy any position. Measurement concerns magnitudes which are, on the contrary, continuous: their parts occupy certain positions and are mutually contiguous.

This situation in Greek mathematics radically changed of course after the discovery of the problem of incommensurability, because of the conflict between commensurable and incommensurable segments, which led to the consideration of irrational numbers and thus became the starting point for the gradual delimitation of the concept of real numbers. Eudoxus' theory of proportions, accepted and advanced in the Fifth Book of Euclid's *Elements*, had been the foundation of this long-lasting effort of mathematicians which was completed only in the nineteenth century, in particular by the work of R. Dedekind, *Was sind und was sollen die Zahlen?* (1888).[2] In contradistinction to the Pythagoreans, Eudoxus produced a purely geometrical theory of numbers which integrated arithmetic and geometry, as well as other mathematical disciplines (among which, in antiquity, one could also fine the theory of music and optics) into a unified theory – the theory of magnitudes. This theory, which arches over the conflict between commensurable and incommensurable magnitudes, is to be regarded as the ancient form of the theory of real numbers.

As a consequence of this development – to which one may also refer in the explanation of why it was only geometry and not arithmetic that became the first axiomatically-constructed theory – there occurred a principal meaning shift in the terms *arithmos* and *megethos*. The extension of the concept of magnitude was, in principle, expanded to all kinds of quantitative notions, not only to geometrical or physical magnitudes (for instance, to time or duration) but also to numbers as such, especially to irrational numbers. The relations within the category of quantity changed as well. Numbers ceased to be independent species of quantity and became a special kind of magnitude. This fact was immediately exhibited in the Latin terminology which managed to operate with the term *quantitas* – sometimes synonymously and at other times, with the differentiated meaning of quantity or magnitude – and *numerus*.

Therefore, if an ancient or medieval author speaks about quantities (magnitudes), he has in mind primarily mathematical objects of a certain kind and not magnitudes in the sense of the theory of measurement, particularly in the context of physical measurement. Let us compare a modern formulation of the so-called Archimedes' axiom in the theory of real numbers –

> for any positive numbers x, however small, and for any numbers y, however large, there exists some non-negative whole number n such that $nx > y$ –

with the original version explicitly ascribed to Eudoxus,

> that of two unequal magnitudes "the greater exceeds the less by such magnitude as, when added to itself, can be made to exceed any assigned magnitude among those which are comparable with it and with one another."[3]

Eudoxus' theory of magnitudes, summarized in the generally-accepted conception according to which mathematics is a theory about magnitudes, lasted, in principle, until the beginning of the twentieth century, regardless of the new ideas advanced in number theory by Dedekind, Weierstrass and Cantor, or in the inquiry into the foundations of mathematics by Frege, Hilbert and Russell. We encounter it, for example, in Newton's *Arithmetica universalis* (1732) where he views numbers as a ratio of a certain magnitude to a magnitude that had been chosen as a unit and is homogeneous with the former, i.e., in the sense of the so-called absolute number,[4] or in Bolzano, who restricts the range of number only to natural or whole numbers and also regards rational fractions as magnitudes.[5] In this almost unified stream, one may obviously find other positions as well, apparently since the reasoning concerning the relations of numbers and magnitudes goes beyond the framework of mathematics as such and concentrates rather on the methodological and philosophical questions of the theory of magnitudes, depending on the problems of measurement.

It seems that one particular exception is J. H. Lambert, who, in his *Architectonics*,[6] on the basis of explicitly formulated criteria, distinguished very precisely the concepts of number and magnitude. According to his view, magnitudes are always represented by a certain named number of the same unit, or they can be reduced to such numbers. At the same time, it is of course not enough to give only the respective numerical value, for we also have to specify its kind, by means of which its qualitative aspects are determined. Besides the fact that every magnitude is characterized by a unit of measurement, it is also distinguished from a number by its dimension and by the specification of a definite unified measuring standard with the help of which it is practically measurable. The unit appears in the magnitude itself; it is of the same kind and corresponds to it in the number of dimensions, too. The notion of a dimension, which was initially interpreted only geometrically, is now generally understood as referring to all species of magnitudes, in order to justify the possibility of measuring heterogeneous magnitudes. In this wider understanding, dimensions then occur everywhere "where the units of different kinds are connected and where numbers of one kind must therefore be connected with numbers of other kinds." Another essential difference between numbers and magnitudes lies in the fact that numbers are discrete, while magnitudes are continuous. For this reason Lambert takes only infinite magnitudes for magnitudes in the proper sense of the term.

These differences are also reflected in the assigned numerical values and therefore they characterize two distinct procedures: measurement and counting. In counting, multitudes and their ratios are represented by whole numbers. The ratios between parts of magnitudes, if we apply a measuring procedure, cannot be expressed by the values of whole numbers. This fundamental distinction does not, however, exclude the use of the arithmetic of whole numbers for the sake of measurement. This follows from the fact that the so-called extensive magnitudes, which are characterized by a measurement unit and by additivity, can be measured in two ways: either "according to number", so far as "we count their parts as wholes", or "according to a degree", if we take into account only that they are continuous. The measurement "according to number" is acceptable only when the resulting numerical values are used for the sake of further calculations and not, however, when the entire procedure terminates in mere counting. Also for the calculations that are something of a connecting link between counting and measuring, one has to require homogeneity of magnitudes, which is guaranteed by the pertinent units and dimensions of measurement.

Under these assumptions, one might even regard – quite contrary to the standard custom – magnitudes as numbers. This possibility becomes a necessity if we measure heterogeneous magnitudes which could not be otherwise directly and immediately compared. Then it is quite inevitable to bring them into mutual ratios which behave as "numbers to numbers". By introducing such ratios, which are not regarded by Lambert as magnitudes in the strict sense of the word, one can measure (indirectly, of course) not the magnitudes themselves, but their ratios.

Lambert's views concerning the nature of magnitudes, their relationship to numbers, as well as measurement and counting, had no influence either on the theory of numbers or on the theory of measurement. The theory of magnitudes had its strong roots in the long-lasting tradition, as well as in an unshakeable trust in the validity of Euclidean geometry. This faith has not been shaken even after the rise of non-Euclidean geometries which proved, among other things, that geometry, because of the conceptual network from which comes the idea of a magnitude, is a hybrid discipline.

Would it not, therefore, be more appropriate if we preserved the conception of magnitudes in the sense of real numbers and introduced the concept of extramathematical magnitudes, which, besides physical magnitudes, would also comprise the magnitudes of other empirical disciplines or, rather, confine the concept of magnitudes to these ranges and, to the contrary, extend the concept of numbers to all types, including complex numbers?

Yet there was still a hesitation concerning this necessary step, even around the year 1900 when specific axiomatic systems within the theory of magnitudes had begun to rise, in spite of familiarity with Hilbert's proof of the consistency of Euclidean geometry by numerical means, as well as with his axiomatization of number theory.[7] Also in this period, students of the foundations of mathematics, under the influence of the paradoxes in set theory, succeeded in justifying the existence of numbers as independent mathematical entities. In this way, they overcame an older position, according to which numbers are only conceptual fixations of magnitudes. It is quite possible that this idea appeared too radical, even to those very people who fought for the replacement of magnitudes by numbers in mathematics. B. Russell, in his work *The Principles of Mathematics* (1903), attempted to construct a mathematical theory of quantities which, on the one hand, would be independent of number theory, yet, on the other, would not aim at meeting the needs of the theory of measurement. In this period one can even witness the culmination of the efforts to demonstrate theoretically that mathematics is a system of magnitudes which comprises geometrical magnitudes, namely, lines, segments, angles, planes and bodies, Eudoxian proportions and all species of numbers, including natural numbers. The adherents of this position attempted to defend it by axiomatizing the theory of quantities, or the theory of the so-called absolute continuous magnitudes.

However, how could one explain the obvious difference in this conception between magnitudes and numbers, especially natural numbers? Yet this question, which simply could not have been ignored, does not get very clear answers. Some authors deny the existence of numbers as mathematical objects and regard a number system only as a conceptual mapping of the system of magnitudes. Magnitudes which are expressed by numbers, obviously non-negative whole numbers, are then rendered as "numerical quantities" (*Zahlgrössen*).[8] Others, again, assume that mathematics contains two different systems of magnitudes: a natural system of ordinal numbers (i.e., the theory of natural numbers) and an artificial system of real and complex numbers; or arithmetic in the narrow sense, namely the arithmetic of whole non-negative numbers, and arithmetic in the wider sense, namely the doctrine of magnitudes that also comprises, besides the theory of real numbers, the theory of geometrical quantities.[9]

In all these cases, whether there is talk about magnitudes in general or about numerical magnitudes which have, in essence, the same meaning as numbers in the current ordinary sense of the term, the construction of theoretical mathematics itself is always at stake, be it the arithmetic of natural or real numbers or geometry (more precisely, mathematical geometry) but not, however, a certain extramathematical theory, for instance, a theory of physical magnitudes.

If we compare the expositions concerning magnitudes and the theory of magnitudes in mathematics with the comprehension of magnitudes from the point of view of the theory of measurement, we cannot overlook the fact that the concept of magnitude has at least two different meanings. To be precise, one has to distinguish the various kinds of geometrical magnitudes, according to whether we interpret them in the context of mathematical or physical geometry. This differentiation can be justified by the claim that the magnitudes of physical geometry need not have precisely the same properties as their corresponding magnitudes of mathematical geometry. So, for example, volume is an additive magnitude from the standpoint of mathematical geometry, yet, if we view it as a magnitude of physical geometry, it need not always satisfy the conditions of additivity. Indeed, as far as we understand the concept of magnitude as an extramathematical notion, we cannot identify it with the concept of numbers. Speaking about magnitudes in the empirical sciences, we certainly do not have in mind only a certain real number.

Hence, if numbers are not a special kind of magnitude, but specific mathematical entities, it is not at all expedient to use the term 'numerical magnitude'. Consequently, the extension of the concept of magnitude in mathematics also contracts and refers only to the magnitudes of mathematical geometry.

3.2. QUANTITIES AND MAGNITUDES

By differentiating between numbers and magnitudes we actually return to the original conception of Greek mathematics. Nevertheless, we do not have to accept its fundamental classificational criterion and assume that magnitudes are continuous, while numbers are discrete. This classification – which is defended today on the grounds that, in measurements, the measured properties are being mapped onto the set of real numbers – is not disjoint, for discrete magnitudes also exist, e.g., electrical charge (the charge of one electron).

At the same time, a question emerges out of these claims as to the relationship between numbers and magnitudes, on the one hand, and the category of quantity, on the other. If we understand *quantity* as a philosophical category, there is no doubt that the concepts of number and magnitude, exactly because they reflect the quantitative aspects of objects, phenomena, and processes of objective reality, belong into this category. This also distinguishes them from other concepts, in particular from the concept of attribute or property, which reflect qualitative properties of these objects, phenomena and processes, and therefore fall under the dual category of *quality*. Should the terms *quality* and *quantity* designate only these philosophical categories, we might concentrate, in the analysis of problems of measurement, only on the investigation of mutual relations between magnitudes and properties. We would not have to wrestle with the problems that follow from the ambiguity of these terms, especially the term *quantity*.

Disregarding the ordinary usage of the term *quantity* in the sense of an *amount*, this term signifies *magnitude*. This variability of meaning is, to a considerable degree, conditioned by purely linguistic reasons. The possibility of conceiving the German word *Grösse* in the sense of 'size' or 'magnitude' leads to such expressions as *die Grösse der Grösse Länge* ('the magnitude of the magnitude of length') and even to confused formulations if we do not distinguish in a sufficiently clear way the predications *is* and *has*.[10] No wonder, therefore, that preference has been given to the synonymous, terminologically differentiated expressions, namely *die Grösse der Quantität Länge* ('the magnitude of the quantity length', analogically, in English usage), even if these two terms are still used with the same meaning, or eventually with different specifications, such as *quantity, quantitative magnitude*, and the like.

By virtue of this terminological shift that has been conditioned quite accidentally by the ambiguity of words (since the Czech language has here two different expressions, *velikost* and *veličina*, there is no such problem in it), changes also arise on conceptual and theoretical levels.

Quantity is understood as anything that can be numerically mapped in some way: every quantifiable – countable or measurable – property. Under a quality one understands those properties, or perhaps relations and the like, which are not measurable. Of course, various views arise with regard to the character of numerical mapping, contingent upon the interpretation of the concept of measurement as to its sense and extension. Qualities, properties, or attributes which can be numerically mapped, can sometimes be taken for quantities, even if the assigned numbers do not express a size (magnitude). A mere linear ordering of the quality involved is admitted as a sufficient condition of the existence of a quantity. However, in most cases, stricter requirements are imposed upon the definition of quantity. In order to be able to say that some property is a quantity, the mapping must apprehend the size (magnitude) too. We are already well acquainted with both of these points of view: the first one corresponds to the wide conception of measurement and the second one refers to the narrow conception. Campbell[11] contracts this narrow delimitation of the concept of quantity even further, so that it is confined only to fundamentally measurable qualities which satisfy the condition of empirical additivity. Consequently he regards every property that is only derivatively measurable (e.g., density) as a quality.

A further, much more significant consequence of the original synonymous use of both of these terms, is their gradual conceptual differentiation on various planes of abstraction, which depends on their specific characteristics and on their diverse connections with other notions too.

If the measurable and non-measurable properties are joined with the philosophical categories of quality and quantity, the emphasis is put on their ontologico-gnosiological aspects. This tendency leads, quite inevitably, either to an *ontological* or to an *anti-ontological* position.

In the first case, the relationship between quantities and magnitudes is transferred from the common conceptual plane into two hierarchically-arranged levels in the direction of a greater concreteness: from a conceptually-reflected plane to an objective one. Thus, quantity constitutes an objective basis for the corresponding magnitude that is conceptually fixed in the system of a definite discipline. Quantity, which encompasses various quantitative aspects of the objectively existing objects, phenomena, and processes, is represented by the same or different magnitudes, depending on the system of measurement units and on the mutual relations between basic magnitudes.

The *ontological* position points out the necessity of analyzing the problems of measurement on different hierarchically-ordered levels. This position emphasizes quite correctly the significance of the ontological level. Without this objective basis we might not be able to justify conclusively why some properties are measurable, why some other properties are not measurable thus far, and, finally, why still other properties are not at all measurable. Yet, due to the identification of the quantitative aspects of objects, phenomena, and processes of objective reality, which are reflected in various magnitudes, with the concept of quantity, the ambiguity of the term *quantity* increases even more. This also prevents the possibility of characterizing the highest level of this hierarchy – the level of philosophical categories.

In the second case, the view of the problems of measurement is also extended, yet their stratification leads in the opposite direction – toward a greater abstractness: from the conceptual level to the theoretical plane, limited, furthermore, only to the language of science. This distinctly *anti-ontological* doctrine is inspired by the familiar subjectively idealist views of the neopositivists. To them, the discernment of qualities and quantities does not mean a difference in the objective reality, but only a differentiation in our conceptual system, in the language of science; it means a distinction between a qualitative language, which contains one-place predicates, and a quantitative language, which is characterized by functions with numerical values. In this conception, the concept of quantity is thus identified with quantitative notions, with metrical predicates. The term *quantity* is only an abbreviation for the term *quantitative concept*.[12]

Owing to the reduction of quantities either to concepts or to linguistic expressions – by means of which operationalism transforms the concept of magnitude into measuring operations and brings this *ad absurdum* by a complete rejection of quantities – measurement loses its objective basis and becomes, if consistently thought over, a merely conventional, subjective affair. We do not consider it proper in this connection to deal in detail with all the controversial and unacceptable consequences of this distinctly one-sided standpoint. Therefore, we shall mention only one consequence of a semantical nature, to which precisely the neopositivists should not acquiesce. If *quantity* is only a simplified expression of the term *metrical concept*, then it can hardly be accepted as correct that quantities (e.g., length in cm) have different numerical values.[13] The assertion that numerical values can immediately be assigned to metrical concepts, for instance, to the *concept* of length, is therefore, obviously, semantically faulty.

The concepts of quantity and magnitude are also differentiated at distinct levels of abstraction. If we set forth this problem and assume that these are different notions, only two possibilities can be considered: either magnitudes are more abstract than quantities or, on the contrary, quantities are more abstract than magnitudes.

In the first case,[14] quantity is a particular, concrete instance of a magnitude. For example, the length of a rod, say, the length of a rod in meters, is a quantity, while *length* is a magnitude. Every quantity (e.g., a certain length, a particular weight in kg, or a certain temperature in °C) assigns to every object in a given area a determinate number that specifies the quantitative value of the respective magnitude – in length, weight and temperature. Thus, if *quantity* is delimited, not only by stating the object and its quantitative aspect but also by stipulating the proper measurement unit, then it is undoubtedly understood as a quantum of something in the sense of a qualitatively-determined magnitude.

In the second case,[15] it is exactly the other way around: *length* is a quantity, while the length of a table, for example, or the length of a table in m, is a magnitude.

From the methodological standpoint, both these conceptions, irrespective of the way they define the relationship between quantities and magnitudes, are entirely congruent. In a similar way they grasp the relationship between concrete, quantitative aspects of objects or phenomena (objectively existing in various degrees and eventually also with a more or less ascertainable size (magnitude) on the empirical plane) and their theoretical generalization in terms of different equivalence classes. Hence, if quantities are held to be concrete expressions of these aspects, magnitudes are equivalence classes of these quantities, or equivalence classes of classes. And, on the contrary, if magnitudes appear to be more concrete entities, then, in turn, quantities are defined with the help of different equivalence classes. This approach, in terms of which the mutual bonds between empirical and theoretical characteristics of an object of measurement are to be expressed, also asserts itself in Helmholtz and Menger's conception. This conception is based on the differentiation of quantities, interpreted as ordered pairs (the first member of which is some empirical object and the second, some number) and on classes of mutually consistent quantities which Menger designates by using Newton's term *fluent* (variable).[16]

The *ontological* and *anti-ontological* standpoints grasp the hierarchical arrangement of different levels on which the object of measurement can be conceptually fixed, only from a definite, one-sided view. By means of a unification and systematization of these conceptions we arrive at a differentiation of at least three, gradually ordered, mutually conditioned levels.

3.3. THE OBJECT OF MEASUREMENT

When we define the object of measurement – an operation which may be performed at various levels and with a distinct measure of abstraction – we must undoubtedly start from its objective basis, which comprises very differentiated species of objects, processes and phenomena of the objective reality, having various properties and entering into different relations. In the process of measurement, these objects as such are only the measured objects, not, however, the objects of measurement.

On the first, *ontologico-gnosiological*, level the object of measurement is delimited by various properties of the measured objects, which we shall regard as their *qualitative* and *quantitative aspects*. Since this dichotomy is obviously too crude, it is necessary, in our view, to distinguish two specific kinds of quantitative aspects.

Under *strongly* quantitative aspects, we shall understand such properties of the measured objects which admit of the discernment, not only of degrees, but also of sizes (magnitudes).

Under *weakly* quantitative aspects, we shall understand those properties which admit only the distinction between degrees.

Qualitative aspects do not satisfy either of these conditions.

On this ontologico-gnosiological level, the object of measurement may be defined relatively separately and independently, in particular with respect to the result of measurement, to the numerical values that are assigned to these aspects.

The differentiation of qualitative, weakly and strongly quantitative aspects, rests, however, not only on the real nature of the various properties of the measured objects which is primary and, in this sense, independent of the manner of how we reflect it conceptually, but also, secondarily, on the attained degree of our knowledge. These aspects, therefore, are not sharply delimited and immutably established once and for all. Between them, there exist various transitions, mutual connections, and relations; their knowledge is important precisely when we set forth the question as to what degree may one also hold qualitative aspects to be measurable in the narrowest sense of the term, or whether and under what conditions qualitative aspects are also measurable.

If we consider the different properties of objects or things, phenomena and processes of objective reality, we presuppose a certain classification of these objects into individuals, properties of individuals, relations between individuals, classes of individuals, properties of classes of individuals, etc., which is philosophically supported by the differentiation of the world into things, properties, and relations.[17] What we regard as a thing and what as a property, what as an individual and what as a class, how we distinguish things and processes – all this is contingent (depending on the dialectics of whole and part, the universal and the individual and the particular) on the nature of the division of reality that is relativized to the level on which natural and social phenomena are investigated in one or another scientific discipline. Regardless of the specification of objects and the ways of their investigation, a thing or an object is, for instance, a forest, an individual tree, or its branch or leaf, just as a molecule, atom, electron, or positron. By making a reasonable decision in favor of some division, whether it be on the macrophysical level, or on the levels specific for the domains of investigation in various scientific disciplines, we also define a set of objects $\boldsymbol{D}$, or $\boldsymbol{E}$, if the given domain refers immediately to the phenomena of the objective reality. In terms of this relativization as to the given domain of investigation, what can be regarded in a given set of objects as a thing, and what can be viewed as a collection of things is also determined; or, to express it in the language of set theory, what can be rendered an individual and what can be regarded as a set of individuals. One must distinguish, likewise, different kinds of properties as well. Some of them will refer only to individuals, others only to sets. It can be said about an individual thing, for example, that it is smooth, heavy, colored, and the like, but it cannot be stated that it is also denumerable. This property refers, not to individual things, but to collections of such things, to sets of individuals, and sets of sets, etc. Some properties will have different specifications in different theories. Thus, for example, relative velocity is additive in classical physics, but not in the special theory of relativity.

Therefore, if we want to move from the first level of objectively existing aspects to a further, *empirico-mathematical* level, we must strictly differentiate between the properties which refer to the measured objects and the properties which can be meaningfully assigned only to collections of these objects. Of course, this is far from sufficient. We must also set forth the question as to which properties are, in fact, relevant for the clarification of the object of measurement in various specifications. Indeed, what can be measured on individual objects and what on the sets of objects? This question can be divided in two parts: which strongly or weakly quantitative aspects refer to sets of objects, provided that these can be constituted on a very general level and that their elements might be almost entirely incomparable objects? We shall answer this question only in our analysis of the properties of counting and measurement (Section 5.2). But which strongly or weakly quantitative aspects can refer to individual objects? Apparently, there is a whole host of these aspects which express some property in different magnitudes (size), or at least in different degrees.

If we are to regard these properties as magnitudes, we must recognize them as individual objects, identify them, compare them, abstract and generalize them so much that we could numerically express them in the process of measurement. This conceptualization, based on a gradual building of equivalence classes of distinct generality, is limited, especially as to its abstractness, by a number of conditions that are specific for a certain way of measurement, as well as for a certain kind of magnitude.

This procedure, which, of course, also encompasses the foregoing level, can be illustrated in a very simplified form by means of the following conceptualization of the length magnitude.

Imagine that we have at our disposal a series of very heterogeneous individual objects (for example, a fence, glass, apple, stick, rose, a bit of dust, hair, number 5), and that we strive to find out whether they have some common property. By mutual comparison, which is neither easy nor always entirely unambiguously realizable, we may, for instance, find out that some of them occupy a certain location and are extended. This aspect, which can be very clearly recognized in some of them, can be called *length*. Let us say that, through primary homogenization of these heterogeneous objects, we have arrived at an equivalence class {fence, stick, chair, hair, glass, apple} which comprises such distinct objects on which we pass judgment that they are equivalent with respect to *extension*. This is only the first step toward the constitution of the *magnitude* of length at which we have discerned its fundamental qualitative aspect by virtue of which, for instance, *fence* undoubtedly differs from the *number 5*, which obviously does not belong in this equivalence class. This still considerably heterogeneous equivalence class can be further homogenized if we consider, not only such objects of which we know from experience that they are extended, but, more narrowly, only individual objects of the same sort, e.g., sticks, beams, rods, fences, or only different rods. Consider, for the sake of simplicity, the equivalence class:

$$\{rod_a, rod_c, rod_d, rod_f, rod_h, rod_j, rod_k, rod_o, rod_p, rod_q, rod_r\}.$$

This class can be further homogenized if we inquire to what degree does the property of extensiveness belong to its elements. Suppose that in our equivalence class one can find the following subclasses of equally long rods, which may, of course, contain only one element:

$$\{rod_a, rod_f, rod_h, rod_o\} \qquad \{rod_c, rod_d, rod_r\}$$
$$\{rod_j\} \quad \{rod_k\} \quad \{rod_p\} \quad \{rod_q\}.$$

Since the elements of these classes can be identified with regard to the property of extensiveness, we obtain a further equivalence class, the elements of which will be equivalence classes of equally long rods, namely equivalence classes of one-element equivalence classes, say

$$\{l_u, l_v, l_w, l_x, l_y, l_z\}.$$

We would have arrived at similar results, of course, had we considered other extended objects.

These already considerably abstract equivalence classes, which characterize length in various degrees, must be made concrete in some way, however, if we are to find out empirically to what degree the measured objects mutually agree or differ in this respect. To every class l^* there corresponds empirically some specimen of an extended object, say, some rod. Any rod, say, rod_f, can be chosen as a model, as a standard rod, and then other rods, or any other objects of which we know that they are extended in some degree, may be directly compared. In this way we find out, of course, only whether a certain rod or another object is equally long, longer or shorter than the rod_f.

The topological characteristic of length enables us to order various extended objects or, on a more abstract level, different degrees of length too, regardless of their reference to concrete objects, into a definite series, according to the relations *more* or *less*. Empirical series of distinctly extended objects can be numerically mapped onto a series of numbers which, owing to the order of its elements, represents only degrees (not, however, magnitudes or sizes of length) so that a larger number corresponds to a greater degree, and a smaller number, to a smaller degree.

We shall regard as a *non-metrical magnitude* such a magnitude which associates a certain qualitative aspect of objectively existing objects, for example, their extension, with a weakly quantitative aspect of a degree in which this property can manifest itself. In the same sense we might, of course, also talk of *topological magnitudes*.

If we are to view some magnitude as a *metrical magnitude* that associates a certain qualitative aspect of objectively-existing objects – for example, their extension – with a strongly quantitative aspect of a magnitude, we must furthermore introduce a measuring unit, for example, by mapping the length of the model rod into a number of a certain size.

By introducing the measuring unit of length, the basic process of comparison, with the help of the model rod, can be materialized by means of counting the number of these units. The length of an object can then be expressed by some number in a proper sense of the term. A sequence of rational numbers then represents the sequence of numerical values which are assigned or can be assigned to the magnitudes of lengths of empirically measured objects. So far, we need not be interested in the concrete application of this way of comparison or in further conditions under which it is realizable.

With the help of this model, which in essence reproduces different historical periods in the conceptualization of a metrical magnitude of length, we can also elucidate the specificity of the object of measurement at the second level, just like the nature of magnitudes, particularly from the structural standpoint.

Since every magnitude associates some qualitative aspect of the measured objects with some weakly or strongly quantitative aspect, we might characterize them on the empirico-mathematical level quite generally by a definite function f (e, n) with an empirical domain of arguments and a numerical domain of values. The arguments of the function f can be interpreted as *empirical variables* and their values as *numerical variables*, i.e., as variables the values of which are numbers.

If we assume that *magnitudes represent the object of measurement on its empirico-mathematical level*, which is fundamental for grasping the essence of measurement, it follows from this that the object of measurement on this second level cannot be conceived any more in isolation from the result of measurement. This dialectical relationship is made concrete, of course, only in the very process of measurement, by virtue of a specification of numerical values which are assigned to empirical variables.

Empirical variables, which represent properties of the measured objects, can be applied as referring to various equivalence classes: either to the measured objects, or to their qualitative aspects, or to qualitative aspects of the measured objects of a definite sort. Each of these admissible interpretations of empirical variables is conditioned by different circumstances.

If it is known that the qualitative aspects belong just to those objects on which, or by means of which, we measure them, or if we have at our disposal a theoretically and operationally, sufficiently corroborated qualitative concept of the respective magnitude, it will be sufficient to relate the empirical variables only to the qualitative aspects of the measured objects. If we measure, for example, the temperature of different objects, today we know very well on what objects we can measure this magnitude and so we shall definitely not assume that it could be ascertained, for instance, with regard to sensations or desires. This interpretation will be utilized, of course, when we consider the measurement of magnitudes in general, regardless of the question, on what objects or by means of what objects we concretely measure them; for example, if we compare the measurement of time with the measurement of speed. In either case, empirical variables will directly refer to the qualitative aspects of the measured objects, i.e., to more abstract equivalence classes.

If the necessary preconditions under which the object of measurement can be analyzed on a higher level of qualitative aspects are not satisfied, empirical variables must be related to more concrete equivalence classes of the measured objects. As a rule, we encounter this situation at the very onset of measurement, when it is not yet fully apparent to what sort of objects the property we are searching for refers, when its scope is not precisely delineated, or if we measure it under new conditions which are not entirely clarified operationally and theoretically. This situation is therefore mainly characteristic of the various cases of extraphysical measurement, when it is often not apparent what magnitude is, in fact, measured, or whether that which is measured really corresponds to our projected goals.

A third possible interpretation is applied if we differentiate the measured magnitude with regard to distinct objects, if we measure it in different boundaries, and if we restrict the measurement procedure only to a certain species of objects or, even more concretely, only to a single object; for instance, if we want to distinguish the measurement of the mass of a billiard ball, an atom, and the Moon.

On the empirico-mathematical level, only one interpretation – the numerical one – can be given to numerical variables, which in themselves represent the so-called scale values of numerical scales, obtained as a result of various measuring procedures. The range of values of this variable consists either of the so-called ordinal numbers, which represent only ordering, or of the so-called cardinal numbers, which also represent magnitudes, namely the ordering of denotata of empirical variables and their size.

Empirical variables thus delimit the qualitative aspects of magnitudes on the most diverse levels of abstractness: the *type* of magnitude (for example, length); *kinds* of magnitudes (for example, distance, height, depth, breadth); its *specifications* (for example, wavelength); or its *concrete instances* (for example, the length of my writing desk).

Numerical variables represent weakly or strongly quantitative aspects of magnitudes; covering either the entire scope, in which the type, kind, or a specification of the given magnitude can be objectively found, or only its concrete instances. Therefore we must take either sets of numbers or their elements for a numerical component of magnitudes. Thus, for example, in the case of a nonmetrical magnitude of hardness, the set of ordinal numbers $\{1, 2, 3, \ldots, 8, 9, 10\}$, in the original Mohs' conception, or another set of numbers $\{0{,}2,\ 0{,}3,\ 0{,}4,\ \ldots,\ 9{,}0,\ 9{,}5,\ 10\}$, in a more refined classification, represent the ordering of materials according to their hardness. The position of rubidium, for example, is then designated in this second classification by the serial number 0,3.

The differentiation between the sets of values of numerical variables and their elements is, of course, also valid with regard to metrical magnitudes, where, in addition, some measurement unit is to be stipulated.

The difference between magnitudes and their values enables us to outline the kernel of the controversial questions concerning the mutual relationship between magnitudes and quantities (Section 3.2).

If we then comprehend magnitudes as sets of ordered pairs $\{\langle e, n\rangle\}$, and each of their concrete numerical values, representing the ordering or size of a certain empirical variable, as an *element* of this set, for instance, $\langle e_i, n_i\rangle$, we need not distinguish between quantities and magnitudes on the empirico-mathematical level of conceptualization of the object of measurement.

In this way we avoid the difficulties which follow from the familiar ambiguity of the term *quantity*. When we begin with the original conception of magnitude and quantity, it is obvious that we find here two different concepts of considerably distinct generality which do not belong in the conceptual network of one and the same discipline.

The concept of magnitude is the fundamental concept of the theory of measurement. The concept of quantity – like its twin, the concept of quality – is a philosophical category. These "stages of distinguishing, i.e., of cognizing the world, focal points in the web, which assist in cognizing and mastering it,"[18] cannot be viewed as primarily relevant notions of the theory of measurement.

From this differentiation there must be inferred the only possible conclusion which is acceptable from the standpoint of Marxist-Leninist philosophy. In order not to confuse, or rather not to subordinate, the concept of quantity under the concept of magnitude – which in its consequences means nothing else but the rejection of the categorial system in philosophy and of its function in the process of cognition of the objective reality, a rejection which suits only the "antimetaphysical" standpoint of neopositivism, pragmatism or instrumentalism – the concepts of *quantity* and *quality* must be understood, quite unambiguously, in just the sense of *philosophical categories*.

We have still to assess the relationship between non-metrical magnitudes and topological concepts and between metrical magnitudes and metrical concepts. This will facilitate our explanation of the third, *theoretical*, level of conceptualization of the object of measurement. On this level, the object of measurement will be represented either by *metrical concepts*, if we accept only the narrow conception of measurement, or by *topological concepts*, if we also include measurement in the wider sense of the term in the scope of measurement. However, when conceptualizing the object of measurement on the theoretical level, we leave out classificational concepts, for, in our view, measurement in the broadest sense of the term cannot be justifiably accepted as measurement proper.

These concepts could be understood as abstraction classes of magnitudes, which are characterized by the same empirical, formal, theoretical, and perhaps also operational properties. Topological concepts are abstraction classes of non-metrical magnitudes; metrical concepts are abstraction classes of metrical magnitudes.

The difference between non-metrical and metrical magnitudes, on one hand, and between topological and metrical concepts, on the other, is sometimes interpreted, however, only as a difference between observational and theoretical concepts, between concrete instances of magnitudes (if we measure, for instance, the length of a table, or the length of the path of a moving body) and the magnitude itself (if we consider, for example, the measurement of length regardless of the fact on which object or with the help of which objects it is measured).

This neopositivist conception of observational and theoretical concepts, based on a strict dichotomy between observational and theoretical languages of science, overlooks the quite apparent fact that magnitudes are neither exclusively observational notions nor exclusively theoretical ones. We have shown that magnitudes constitute a dialectical unity of qualitative and quantitative aspects of really existing objects, phenomena, and processes, which is conceptually reflected in different equivalence classes. In conformity with the manner of interpretation of empirical variables at the empirico-mathematical level, it is only the proportion between their observable and conceptual characteristics that is changing. The type of the magnitude is, obviously, more "theoretical" than its concrete instances; yet, as much as we like to talk in both cases about magnitudes, the "observational" component cannot be completely eliminated in the type of magnitude either. To take magnitudes – the type of magnitude or its different sorts – for theoretical concepts means nothing else but to identify them with topological or metrical notions.[19]

In principle, one cannot object to the possibility of abstracting from the observational characteristics of magnitudes. The completion of the gradual conceptualization of the object of measurement indeed quite necessarily requires the analysis and synthesis of the common properties of magnitudes; from the standpoint of their relationship to the entire conceptual and theoretical construction of the respective scientific domain, as well as from the point of view of a general theory of measurement. The differentiation of topological and metrical concepts on the theoretical level of conceptualization of the object of measurement is undoubtedly a very significant and valuable affair in the methodological sense. Of course, this conception is acceptable only if the distinction is drawn between particular levels and if respect is being paid to mutual relations between *topological and metrical concepts at the theoretical level and non-metrical and metrical magnitudes at the empirico-mathematical level*, as well as between *qualitative, weakly and strongly quantitative aspects at the ontologico-gnosiological level*.

3.4. MEASUREMENT UNITS, NAMING AND DIMENSION

It follows from the characteristics of magnitudes as functions with empirical arguments and numerical values, that every magnitude can be expressed by some *named number*. Naming refers to the empirical variables that characterize the qualitative component of magnitudes, while numbers represent their quantitative determinations: in the case of non-metrical magnitudes only ordering is represented, while in the case of metrical magnitudes, also their size.

The nominal component of metrical magnitudes, represented by named cardinal numbers, is identified with so-called dimension, especially in the theory of physical measurement, while the numerical component is interpreted as a set of multiples or portions of the measurement unit which is homogeneous with the given magnitude. It is to be assumed for non-metrical magnitudes, at least *per analogiam*, that every element of a series of numerical values in the ordering that is specific for each of them, refers unambiguously to a certain degree of the magnitude in question, hence, that the series of numerical values representing these weakly quantitative aspects is constituted by named ordinal numbers. Thus, for example, in Mohs' scale of hardness, the ordinal number 8 designates topaz, while on the Beaufort scale of wind force, it designates a stormy wind (fresh gale) between 55 and 65 km/h. The ordinal number 1 on Mohs' scale, more precisely, *hardness of degree 1*, which represents sodium chloride, and the ordinal number 1 on the Beaufort scale, more precisely, the *force of wind of degree 1*, representing a breeze (light air), are two different named numbers which refer to distinct empirical variables. Therefore, they should not be identified, just as one should not identify named numbers 4 m and 4 kg which represent two equal sizes of qualitatively distinct metrical magnitudes.

Of course, an objection can be raised against this conception of named numbers to the effect that, except in the domain of theoretical arithmetic, we never operate with unnamed numbers, and, likewise, that we never compare, for instance, 4 tables and 3 flower pots. However, there exists a considerable difference between the naming of magnitudes and these named numbers. Naming is a necessary component of every definition of a magnitude, and as such it is inevitable for the differentiation of various qualitatively determined metrical and non-metrical magnitudes. Named numbers, referring to collections of objects, express, regardless of their very heterogeneous nature, the only property common to all sets – their cardinality. The fact that we can compare these named numbers, even to add them, is precisely due to this property of sets. Imagine only that we want to determine the number of objects on a writing desk; the assertion that 5 books + 2 ashtrays + 1 telephone + 1 calendar = 9 objects, is evidently meaningful.

Presupposing, therefore, that the numerical component of non-metrical magnitudes composes a sequence of named ordinal numbers, we might regard any of them even as a *unit of ordering* of particular degrees of their qualitative components. This generalization of the concept of *measurement unit* which is, of course, quite unusual, could be justified in our conception by taking non-metrical magnitudes to be a specific kind of magnitude.

Perhaps we do not have to emphasize that the units of ordering of non-metrical magnitudes differ from the units of measurement of metrical magnitudes, just as measurement in the wider sense differs from measurement in the narrow sense. Let us mention at least one essential difference. If we choose some measurement unit for a continuous metrical magnitude, i.e., a certain size of the measured magnitude, as the basis for comparison, then for objective reasons, every further cardinal number, representing strongly quantitative aspects of this magnitude, will already be unambiguously determined. The choice of units of ordering is, of course, entirely arbitrary and it restricts the selection of further ordinal numbers only by a condition that each of them must be greater or smaller. For example, if we choose the ordinal number n for the representation of the ith degree of a non-metrical magnitude, we can select any ordinal number m such that $m > n$ for the representation of the $i + 1$ degree and, likewise, any ordinal number k such that $k < n$ for the representation of the $i - 1$ degree. As a rule, for the sake of a greater lucidity, the unit of ordering is identified with the beginning of the ordering, for example, with the ordinal number 1, or perhaps even with the number 0, and all other ordinal numbers j are chosen so that they meet the condition $j > 1$. The unit of ordering then represents the smallest degree of a non-metrical magnitude, for example, the least desirable alternative in the case of ordinal utility.

For metrical magnitudes, of which physical magnitudes are the paradigm, the unity of qualitative and strongly quantitative aspects manifests itself in the delimitation of the magnitude itself, as well as in its specific instance – in the unit of measurement. Since the numerical component of the measurement unit can be put as equal to 1, the measurement unit is a certain magnitude of a basic size.

The naming of a measurement unit refers, not to a type or kind of magnitude, but rather to its concretization, for example, to the length of some standard object, say, of a model rod. At the beginning, the numerical value of this specific magnitude need not be explicitly stated. It is completely sufficient to know from practice that the measurement unit in its objectivized form – as a *material* measurement unit – represents for us a definite size of the measured magnitude and that it also facilitates comparison with objects on which this magnitude is exhibited in different sizes.

The measurement unit itself can be understood, in the *conceptual* sense, as an equivalence class of objects having the same size – with regard to the property considered – as the standard object, the material measurement unit. Thus, for example, 1 kg is an equivalence class of all objects having the same mass as a certain model object that is deposited at the International Bureau of Weights and Measures in Sèvres. Kilogram is, then, the name of this equivalence class. In the same way, for example, pound, ounce, and the like, are names of some units of mass measurement, if we use some other equivalence class. We measure metrical magnitudes (for instance, length) in multiples and portions of a definite unit of measurement, for example, in cm, dm, km, and the like, previously also by using various other measurement units, such as ells, double paces, feet, yards, miles, fathoms, acres, etc.

If we can choose the numerical component of the measurement unit as equal to 1, a qualitative aspect predominates in its specification, regardless of the fact that we can operate with it as with a number and perform with it various arithmetical operations. In reality, measurement units, their multiples and portions, the so-called multiplicational or divisional measurement units, are not numbers in the strict sense of the term, even if we calculate with them as with numbers, abstract from their qualitative components and assume that numerical values of magnitudes are expressed in qualitatively as well as numerically homogeneous units. It is obvious, *ex definitione*, to regard measurement units as specific instances of metrical magnitudes.

Therefore, it is not correct when any magnitude X in physics [20] is generally expressed as a product of a numerical value $\{X\}$ and a measurement unit $[X]$ according to the relation

$$X = \{X\} \cdot [X],$$

or when the measured value of a magnitude is more concretely held to be a product of a measurement number and a measurement unit. An expression of the form

$$\text{magnitude} = \{\text{numerical value}\} \cdot [\text{unit}]$$

defines a magnitude, rather inexactly, as a certain function existing between a set of values of empirical variables and a set of values of numerical variables, which also represent, due to the measurement unit, their sizes. In the second case, an analogical expression, namely

> the value of a magnitude = measurement number · measurement unit,

characterizes this function for a certain value of the magnitude. Both these expressions might be conceived rather as symbolical expressions of metrical statements – either generally, for example, "Mass is being measured in kg", or concretely, for instance, "The mass of the body A is 76 kg" – or of the procedure for calculating numerical values of magnitude. However, in no way could one consider these expressions as definitions of the concept of magnitude, or of the value of a magnitude, which would contain in the definiens the numerical operation of multiplying the measurement number and the measurement unit.

Our objection against this quite ordinary formulation holds even more for the so-called dimensional formulas of dimensional analysis,[21] which characterize in a symbolized form either magnitudes derived by means of the *products of powers of basic magnitudes*, or the measurement units of these secondary magnitudes as the *products of powers of the basic measurement units*. Obviously, the numerical operations of multiplication and exponentiation only refer, in this case, to the numerical values of a magnitude or measurement units and not, however, to metrical magnitudes or measurement units themselves.

Had we interpreted the above-mentioned formulations, *verbis expressis*, from a theoretical and methodological points of view, the contrary position would have conflicted with the qualitative character of the numerical laws that reflect the objective law-like regularities of the motion of matter, as well as with the invariance of functional relations between magnitudes in the formulations of these laws, which is independent of the manner of the numerical representation of their size.

Insofar as we encounter in physics such definitions of derived magnitudes as, for example,

> force = mass · acceleration,

and an equation of the type

$$F = ma, \qquad (3.4\text{–}1)$$

we must be well aware that here we are dealing only with symbolized expressions and not with definitions or numerical equations: the symbols F, m, a designate neither numbers nor numerical variables but the names of the respective magnitudes. The operation of multiplication is not a definitional sign of the magnitude of force, but designates only a certain functional relation.

In the given case, what is involved is an *illustrative* expression of a consequence of Newton's second law of motion: "The acceleration of a body is directly proportional to the force, and indirectly proportional to the mass of the body." According to this law of force, the acceleration of a body is thus greater the greater the force and the smaller the mass of the body. This functional relation can be expressed by the formula

$$a = F/m, \qquad (3.4\text{–}2)$$

in which a designates acceleration, F force and m mass. Therefore, if we know a definite numerical value of the force F, with which we act on a given body, and a certain numerical value of its mass m, we can calculate, in accordance with formula (3.4–2), the numerical value of the acceleration a as their ratio.

The consequence (formula (3.4–1)) expresses a similar functional relation: force is directly proportional to the mass of the body and acceleration. If we know, therefore, the numerical value of the mass m and the numerical value of acceleration a, with which the given body is put into motion, we can establish the numerical value of the force F, with which we have to act on, by multiplying these numerical values. At the same time, of course, we always assume that the measured numbers of the magnitudes considered are expressed with the help of units of measurement of a definite coherent system.

Hence, if we talk about a product, a quotient, and the like, these numerical operations refer only to the numerical values of magnitudes or measurement units. It is possible to abstract from a naming or a dimension of metrical magnitudes and to consider only their size, represented by certain cardinal numbers, only because we know in advance the functional relations between the magnitudes involved and because we simultaneously presuppose a definite system of basic and derived units of measurement.

In practice, of course, every physicist knows this very well. Yet in theory, such an inexact manner of expression is inadmissible, for it may very easily lead to a misinterpretation of the function of physical laws, of the concept of magnitude, and their relationships with numbers, etc.

These incorrect consequences are particularly manifested in the so-called formal theory of measurement which is defended by many methodologists of behavioral and social measurement.[22] When these methodologists delimit the structure of magnitudes, they also count dimensionless real numbers among metrical magnitudes and introduce as a primitive notion of this structure a certain operation of multiplication about which they assume that it has the same formal properties (closure, commutativity and associativity) as the numerical operation of multiplication. Yet they do not explain what makes the conceptual difference between these structurally conforming operations. For various reasons, this purely formal comprehension of magnitudes is unacceptable: it ignores the qualitative difference between numbers and metrical magnitudes, obliterates the difference between mathematics and empirical disciplines, distorts the nature of numerical operations and is, above all, a typical expression of a neopositivistic reductionism.

If we can, strictly speaking, apply numerical operations only to numerical values of magnitudes, and thus only to the size of the measurement units and not, however, to the magnitudes and measurement units themselves, in this sense we must also make the common formulations of the law of invariance more precise with respect to a change of the unit, and with respect to the law of conversion of the unit size, which immediately follows from the former.[23]

According to the *law of invariance with respect to a change of the unit*, or more precisely, with regard to a change of the unit size, a numerical value of a certain magnitude is independent of the units, more precisely, of units of a certain size, by means of which we measure that magnitude. This law rests on the presupposition of the objective existence of properties and relations of real objects, which are independent of their conceptualization; in the given case, on the selection of the size of the measurement unit, either in the same or in different systems of measurement units.

If we start with the assumption of the objective nature of the strongly quantitative aspects of metrical magnitudes and compare two distinct numerical values of a magnitude of the same name, the ratio of these numerical values, expressed by a certain unnamed number, will be constant. The ratio of the size of two magnitudes has thus an "absolute" meaning which is independent of the size of the chosen measurement unit.

This invariance is immediately evident in the expression of numerical values by means of homogeneous units of the same magnitude. For homogeneous magnitudes, so far as they are measured by units of the same size, we might, in general, express the invariance with respect to a change of the unit as a constant ratio of measurement numbers.

$$n/n' = K, \tag{3.4–3}$$

or, more explicitly,

$$kj/k'j = K, \tag{3.4–4}$$

where k and k' are measurement numbers and j designates a unit of the same size.

Essentially every numerical value of a metrical magnitude can be understood as a ratio between the measurement number and the basic size of the measurement unit 1

$$kj/1j = K. \tag{3.4–5}$$

The invariance based on the same ratio of the identically named numbers is, of course, trivial. Much more significant is the invariance of numerical values of homogeneous magnitudes which are measured by means of units of different size, since this enables us to compare not only the ratios of two numerical values but also, mutually, the numerical values of metrical magnitudes. For this case, the law of invariance can be formulated – in accord with (3.4–4) – in the form

$$kj/k'j = hj'/h'j' = K, \tag{3.4–6}$$

where k, k', h, h' are measurement numbers and j, j' are units of different size. From the formula (3.4–6) there immediately follows the equality of numerical values of metrical magnitudes that are measured in units of different size

$$kj = hj', \tag{3.4–7}$$

or, respectively,

$$k'j' = h'j'. \tag{3.4–8}$$

From these equations, there follows, obviously, the equation of ratios expressed invariantly in units of the same size.

From equation (3.4–7) – analogically, also from equation (3.4–8) – the *law of conversion of the unit size* might then be derived in the form

$$h = k(j/j'), \tag{3.4–9}$$

according to which the following holds: if the numerical value k of some magnitude is expressed in terms of units of the size j, then the numerical value h of this magnitude, expressed in terms of the units of size j', will be equal to the k-multiple of the factor (j/j'). This factor, which enables us to transfer numerical values of magnitudes that are expressed in terms of units of one measurement system into numerical values of measurement by means of units of another system, or to mutually transfer basic, multiplicational or divisional, main or subsidiary units in the same system, is called a *conversion* (transformation) *factor*. The conversion factor, for example, for converting the values of magnitudes in centimeters into values in meters is equal 1/100; the factor for the conversion of seconds into minutes is equal 1/60, and so on.

In general, it can be said that, according to this law, the numerical value of a metrical magnitude is *inversely proportional* to the size of the measurement unit. If we change the size of a basic unit by some multiple p, we must change every numerical value by its inverse value $1/p$; and, on the contrary, if we change a numerical value of a given magnitude by some multiple, the size of the respective measurement unit must also be changed in an inverse ratio. Therefore, if we choose a *large unit* for the measurement of some magnitude, its numerical value will be expressed by a smaller number; however, if we select a *smaller unit*, then it will be represented by a greater number.

Both these laws hold primarily for the measurement units of fundamental magnitudes. Since derived magnitudes are defined with the help of these – and this holds likewise for the relationship between basic and derived measurement units as well – these laws can also be indirectly applied to the former and, thus, to all metrical magnitudes.

From the viewpoint of the invariance of functional relations between magnitudes and the objectivity of quantitative aspects of magnitudes, which constitutes the precondition of both these laws, one may also assess the frequently repeated claims to the effect that measurement units of fundamental magnitudes are arbitrarily stipulated, that their choice is purely conventional, merely a question of practical expediency which has no theoretical meaning,[24] and the like. Yet, to what extent is this position warranted, even if it is justified by various arguments, in particular by the non-existence of an objective unit size in the case of continuous metrical magnitudes, by the arbitrariness of the material measurement unit or by the variability of different measurement systems? Obviously, this depends upon our interpretation of the notion of convention – as a justified selection, or as a purely subjective choice?

Measurement units of fundamental magnitudes, insofar as they represent a continuous phenomenon, are indeed conventionally, yet not arbitrarily, selected. Their selection is an expression of a condensed human practice, theoretical considerations and a very diligent process of standardization. In the choice of measurement units, conventionality refers, not to their qualitative, but only to their strongly quantitative aspects: *only the size of basic measurement units is conventionally selected*. This is conditioned by the fact that for continuous magnitudes – in contradistinction to discrete magnitudes – one cannot find further indivisible parts that would unambiguously characterize some unit size. For this reason, these "artificial" measurement units, the application of which is still prevalent, are sometimes distinguished from the so-called "natural" units of certain magnitudes in quantum physics or atomic physics, among which are also included basic physical constants, such as the gravitational constant and Planck's constant.

The constitution of measurement units is preceded, in historical and systematic respects, by practically performed operations of measurement with the help of standard measuring instruments which must be *constant*, *easily reproducible* and suitably *precise*. If these conditions are not satisfied, we could not use them practically as an objective means of comparison. These conditions are also taken into account when we define measurement units of fundamental magnitudes, depending upon the objectively reproducible natural processes and phenomena, for example, upon the wavelength of some light source, upon the period of rotation of the Earth around the Sun, and the like; and also when we gradually make more precise the size of basic measurement units and their standardization.

Let us consider just the entire history of measurement units for measuring length by means of various material standards (for example, feet, inches, ells) up to the original definition of the concept of a meter as 1/40 000 000 part of the length of the Parisian meridian; thence to the construction of a standard meter and to an older definition of the corresponding measurement unit as "the distance between two notches on the international prototype of a meter, deposited at the International Bureau of Weights and Measures in Sèvres, measured at the temperature 0°C and at the pressure 1.01325×10^5 of pascals", and, finally, to the newest, conventionally more precise definition, according to which a meter is the length "equal to a 1 650 763.73 multiple of the wavelength of radiation propagated in a vacuum, which belongs to the transition between the energy levels $2p_{10}$ and $5d_5$ of the krypton^{86} atom."

Although this definition of the concept of a meter has been agreed upon at the General Conference for Weights and Measures, its conventionality is neither arbitrary nor an exclusively practical affair, but a consequence of highly demanding theoretical conceptions.[25] If we could arbitrarily introduce various measurement units, it would not be necessary to distinguish between measurements in the wide and narrow senses of the term, nor between non-metrical and metrical magnitudes. The theory of extraphysical measurement would not face the principal problem of how to defend an extension of measurement without the existence of a, *perhaps indeed only arbitrarily chosen*, measurement unit. Apparently, the size of the magnitude we select as the measurement unit is not so decisive as the fact that it *can* be chosen as the basis for comparing strongly quantitative aspects in order to be able, with its help, to conceive the assigned numbers as cardinal. Before we introduce a measurement unit, we must know what we actually want to measure and whether we can measure it at all.

The choice of basic measurement units depends, not only on the possibility of their objectivization with the help of some material standard, but also on the clarification of the theoretical relationship between fundamental and derived magnitudes. For practical reasons, also for the measurement units of derived magnitudes, one can construct *ex post* different material measurement units; theoretically, however, this is superfluous, since these magnitudes – just like their measurement units – are, in principle, defined concepts. The questions about which magnitudes are fundamental and which are derived, which measurement units may be viewed as primitive and which as defined, cannot be answered arbitrarily. It is evident that we cannot measure the area before we are able to measure the length; or that we cannot measure speed if we do not have at our disposal the units for measuring length and time. Therefore, in measuring different magnitudes, we must inevitably respect the fact that some of them are objectively simpler and can thus be measured directly, regardless of measuring other magnitudes, while others are more complex and already presuppose measurements of the more simply measurable magnitudes.

The choice of measurement units depends, therefore, on *objective relations* between different magnitudes which are theoretically reflected in a twofold way: in the system of fundamental and derived magnitudes and in the system of respective measurement units.

The constitution of a system of magnitudes is a complicated historical process of the cognition of the law-like regularities in objective reality, a process which naturally proceeds from the lower forms of motion of matter toward higher forms. The process of cognition from the simple to the more complex is directly manifested in the development and theoretical systematization of the sciences by virtue of arranging "these forms of motion according to their inherent sequence", and through the classification of the sciences, "each of which analyzes a single form of motion, or a series of forms of motion that belong together and pass into one another."[26] This procedure is also characteristic for each individual scientific discipline. It is therefore no accident that, historically and systematically, the first constituted discipline in physics was kinematics and only later dynamics, and that it has been mechanics, comprising these partial disciplines into one whole, which has become the basis on which further physical domains, for example, the theory of heat, could be built. This is precisely the reason why different sizes of the magnitudes of length, time, and mass have become basic physical measurement units.

Measurement units for these fundamental magnitudes can be encountered in every coherent, theoretically significant, and practically suitable system of measurement units: in the three-dimensional system CGS – *centimeter, gram, second* – or in the seven-dimensional international system – *meter, kilogram, second, kelvin, candela, mol*, etc. If we limit ourselves only to these or other decadic systems,[27] which are mutually convertible in an easy manner, it is not at all decisive, for theoretical or practical reasons, whether we select *centimeter* or *meter* as the basic unit of length, or whether we choose *gram* or *kilogram* for the basic unit of mass.

Only in this sense can the choice of measurement units be considered arbitrary. The conventional nature of this choice is, of course, greater than it is in the case of using synonymous words that have exactly the same meaning in the same language, for instance, sameness and identity, or in different languages, for example, dog, *sobaka, Hund, pes*, and *chien*. Evidently, in every coherent measurement system the main units of derived magnitudes are always unambiguously determined by the selection of basic units of fundamental magnitudes.

Taking into account that basic units can be viewed as the initial concepts of a measurement system and main units as the defined notions, the arbitrary character of this choice of basic units is exactly the same as in the selection of primitive notions in various equivalent axiomatic systems of one and the same theory. Of course, if we compare axiomatic systems with collections of magnitudes instead, i.e., if we consider fundamental magnitudes as the initial concepts and derived magnitudes as the defined ones, the measure of conventionality in the axiomatic systems is, in fact, much greater, because it is not naturally limited by the existence of objective functional relations which, of necessity, must be respected in defining derived magnitudes.

An analysis of these relations, which determine a coherent construction of a measurement system, therefore, must be offered before we are able to determine the size of main units by means of basic ones, of which we ought to assume that they are mutually independent, and thus irreducible. As a rule, the analysis is executed by finding out the dependencies between the dimension of derived and fundamental magnitudes. Knowledge of these then facilitates a conversion of the dimension of derived magnitudes into the "product of powers" of the basic dimensions of fundamental magnitudes. In this analysis, one begins with the requirement that magnitudes of the same dimension must be measured by the same units. However, this is possible only when different names of magnitudes can be reduced to a certain basic or derived dimension. Although this presupposition is desirable for practical reasons, it is not generally realizable – not only for historical reasons, which one usually admitted, but also for theoretical ones.

According to our view, one should distinguish the concepts of naming and dimension. The concept of a *dimension*, which has already been discussed, is understood in various similar meanings. This concept is conceived either as the expression of the relation between simple (fundamental) and complex (derived) magnitudes (according to J. C. Maxwell); or (according to J. Fourier) as the expression of the relation of numerical values of derived magnitudes and the numerical values of fundamental magnitudes with regard to a change of the size of measurement units; or, finally, as the expression of the dependence of derived measurement units on the basic units of fundamental magnitudes in the sense of Helmholtz.

Theoretically, these conceptions are unified either by emphasizing qualitative aspects of metrical magnitudes – and then the concept of dimension can be brought into connection with the concept of naming – or by stressing instead their strongly quantitative aspects, and then a dimension is comprehended as a numerical representation of the size of magnitudes or of their measurement units. Practically, the concept of dimension is then applied in the *dimensional analysis* for the sake of formulating dependencies between derived and fundamental magnitudes, or between main and basic units within the same measurement system, or in the conversion of measurement units from one system to another. In either case, it is assumed that derived magnitudes, defined with the help of fundamental magnitudes, are of the same kind and, hence, of the same dimension, just like the measurement units corresponding to them, which – inasmuch as it is at all possible – are to be named in the same manner.

In every case, according to our judgment, the *qualitative specification* should be emphasized in the context of the concept of dimension. This specification definitely cannot be regarded as a purely quantitative component of metrical magnitudes.[28] Equally incorrect, as we presuppose, is the identification of this notion with the concept of naming.

In the theory of dimensional analysis, dimension is quite generally expressed by the symbol of the respective magnitude in brackets, for example, [v] for velocity, [a] for acceleration, and [F] for force. In order to distinguish between the situation when we want to express the dimension of a derived magnitude itself and the situation when we are interested instead in grasping the dimension of the corresponding measurement unit, two different dimensional formulas are used in the dimensional analysis. The dimension of derived magnitudes is expressed with the help of dimensional symbols of the fundamental magnitudes L (length), M (mass), T (time), etc.,[29] for example, the dimension of pressure is illustrated by the formula

$$[p] = L^{-1} MT^{-2}.$$

The dimension of a measurement unit is illustrated by means of the symbols of basic units of the given measurement system, for example,

$$[p] = m^{-1} kgs^{-2},$$

and this also corresponds to the definition of the main measurement unit of this magnitude: the Pascal (Pa) is a pressure produced by the force of one Newton, uniformly spread over the area with the area content of 1 m^2, which is perpendicular to the direction of the force.

In order to clarify the principle of dimensional analysis, we shall utilize the foregoing discussions. Starting with equation (3.4–1), we shall easily see that only one symbol of a fundamental magnitude occurs in it, namely m (mass). Therefore, we must further investigate in what functional relation is the magnitude a (acceleration) to the fundamental magnitudes of length and time. Since the velocity v of a uniformly accelerated motion is directly proportional to time, the equation

$$v = at \qquad (3.4\text{–}10)$$

holds, and it immediately follows from it that

$$a = v/t. \qquad (3.4\text{–}11)$$

Although a further symbol of a fundamental magnitude occurs in this equation, this is not yet sufficient, unless we ascertain the functional relation between the magnitude v and the magnitudes of time and length. With regard to the fact that velocity is determined by the distance s which a body travels in time t, there holds an equation

$$v = s/t. \tag{3.4–12}$$

The distance s is regarded by us as a kind of magnitude of the type of length l. The following dimensional formula may therefore be stated for the velocity v

$$[\mathrm{v}] = \mathrm{L/T} = \mathrm{LT}^{-1}. \tag{3.4–13}$$

With the help of this formula, we can express the dimension of the magnitude a

$$[\mathrm{a}] = \mathrm{LT}^{-1}/\mathrm{T} = \mathrm{LT}^{-2} \tag{3.4–14}$$

and also the dimension of the magnitude F

$$[\mathrm{F}] = \mathrm{LMT}^{-2} \tag{3.4–15}$$

written in the customary lexicographical arrangement (instead of MLT^{-2}). In the SI measurement system, the dimension of the unit of force is then expressed by the dimensional formula

$$[\mathrm{F}] = \mathrm{m\ kg\ s}^{-2}, \tag{3.4–16}$$

which also corresponds to the definition of the main measurement unit of force:[30] Newton (N) is a force which imparts to a body with mass equal to 1 kg an acceleration 1 $\mathrm{m/s}^2$. In the CGS system the dimensional formula (3.4–15) would, of course, remain without any change. Instead of the formula (3.4–16), however, we would have to use another formula for the dimension of the respective measurement unit, the formula

$$[\mathrm{F}] = \mathrm{cm\ g\ s}^{-2}, \tag{3.4–17}$$

and also to change the name of the measurement unit. In the CGS measurement system, the main unit of force is 1 dyne. The conversion factor between both measurement systems is 10^5, so that

$$1 \text{ dyne} = 10^{-5} \text{ N, respectively, } 1 \text{ N} = 10^5 \text{ dyne}.$$

Therefore, if we know the dimensional formula of some derived magnitude, for example, force, and if we also know in what measurement system its unit is expressed, then we can immediately specify the dimension of its main unit. Naturally, the converse of this relation holds too. At the same time, dimensional formulas enable us to convert every numerical value of the measured magnitude from one measurement system to another. *The dimension of derived magnitudes is thus invariant with regard to distinct systems of measurement units*, if their main units are coherently defined by means of basic units of the same basic dimension. The dimensional invariance of derived magnitudes is an immediate consequence of the dimensional invariance of physical laws, which is conditioned by the fact that functional relations between physical magnitudes described by these laws do not depend upon the manner in which they are numerically mapped with the help of measurement units.

The concept of dimension, however, is *relativized* with respect to a certain system of fundamental magnitudes and, by virtue of this feature, it already differs from naming which refers to empirical variables of metrical magnitudes in an entirely general manner. The *naming* of magnitudes is primary and independent of our way of its dimensional characterization within the dimensional analysis, which is always secondary. Having at our disposal two systems with distinct fundamental magnitudes, we would arrive, in the dimensional analysis for one and the same derived magnitude, at different dimensional formulas, and thus at distinct main units, though we might otherwise continue to use the same name for it. For example, restricting the system of mechanical magnitudes to length and time only,[31] we must regard mass as a derived magnitude with the dimension

$$[m] = M = L^3T^{-2}. \qquad (3.4\text{–}18)$$

Consequently, we would have to change the dimension of other derived magnitudes, for example, for force we would then obtain the dimensional formula

$$[F] = L^4T^{-4}, \qquad (3.4\text{–}19)$$

as well as the definitions of their measurement units. In distinct systems of fundamental magnitudes, dimensional formulas of the magnitude of the same name will be expressed in a different way: they will contain another number or kind of independent fundamental magnitudes and also have a distinct form. In the system of fundamental magnitudes *length, mass, time*, and *electrical current* with the basic dimensions LMTI, which is contained in the system of magnitudes that corresponds to the SI measurement system, and in the system of fundamental magnitudes *length, mass, time* and *electrical current* with the basic dimensions LMTQ, identical magnitudes have a different dimension. Thus, for example, intensity of an electric field is expressed by the dimensional formulas $LMT^{-2}Q^{-1}$ and $LMT^{-3}I^{-1}$, or electrical polarization by the formulas $L^{-2}Q$ and $L^{-2}TI$. Obviously, to the contrary, different magnitudes are expressed by similar dimensional formulas, for example, electric potential by the formula $L^2MT^{-2}Q^{-1}$ and magnetic induction flux by the formula $L^2MT^{-2}I^{-1}$.

The necessity of differentiating between a name and a dimension of magnitudes and measurement units can be justified by additional arguments as well.

The basic dimension L refers to magnitudes of very distinct names: not only to length, but also to height, breadth, depth, path, distance, focal distance, the arm of rolling friction, and the like. These magnitudes, however, grasp the various qualitative aspects of objectively existing objects and phenomena and, from a theoretical standpoint, they have different meanings, too.

In the SI system there exist derived magnitudes of various names which are also measured by different main units, yet have the same dimension. Thus, for example, the momentum of force, which is measured in newton-meters, and heat, work and energy, which are measured in Joules, have a common dimension L^2MT^{-2}. The difference between these magnitudes[32] rests on the fact that heat, work, and energy are scalar magnitudes, while the momentum of force is a vector magnitude. *Dimensional formulas, therefore, do not grasp the essential difference between scalars and vectors.*

If we apply dimensional formulas to so-called proportional magnitudes, i.e., to magnitudes which are expressed by a ratio of two magnitudes of the same dimension (for example, to proportional elongation, proportional permeability, proportional molar mass, and the like), to coefficients (for instance, to a co-product of sliding friction) and, in particular, to plane and solid angles, we shall find out that, in fact, they have no dimension. Since it is valid for an arbitrary plane angle α that it is a ratio of two lengths of the same size l_i/l_i, its dimension can be expressed by the dimensional formula

$$[\alpha] = \frac{L}{L} = L^0 \tag{3.4–20}$$

or, if the symbol L is understood as a number (not very suitably, as we have already mentioned), then by a dimensional formula

$$[\alpha] = \frac{L}{L} = 1, \tag{3.4–21}$$

Strictly speaking, this formula refers to the dimension of the respective measurement unit, i.e., to its definite size.

A plane angle – just like all other proportional magnitudes – is then regarded as a dimensionless magnitude and it is assumed about the dimension of its measurement unit that it is equal to 1. A conclusion is sometimes inferred from this assumption, to the effect that these dimensionless magnitudes are mere numbers.[33] The differentiation of magnitudes into *dimensional* or named and *dimensionless* is customarily explained by the fact that numerical values of dimensional magnitudes depend on a certain system of measurement units, while the numerical values of dimensionless magnitudes are independent of them.[34]

The existence of dimensionless magnitudes and their measurement units with the dimension 1 has always signalized a serious problem for the whole theory of dimensional analysis. Evidently, there follows from the definition of dimensionless magnitudes that for all these magnitudes a single measurement unit might actually be stipulated, namely, the unnamed number 1, and this stipulation could even be independent of every system of metrical magnitudes. However, this numerical value can also be obtained for every ratio of two equally large numbers. Therefore, does it make any sense to talk about measurement units of dimensionless magnitudes if this is always only the unnamed number 1? Thus, if the number 1 is the measurement unit of all dimensionless magnitudes and also of all numerical ratios having the same numerator and denominator, it is to be conceded, at any rate, that all these entities have the same dimension or the same name and, thus, that they refer to the same qualitative aspects of objects, phenomena, and processes of the objective reality.

Yet we cannot avoid a similar, equally unpleasant consequence if we adopt the opposite position and suppose, therefore, that dimensionless magnitudes have no measurement unit in the right sense of the term. This alternative is again in conflict with the fact that different measurement units are used for angles. For the magnitudes of a plane and a solid angle, radian (rad) and steradian (sr) are defined in the SI system as complementary measurement units. Radian is defined as a "solid angle bounded by two radial semirays which intercept on a circle an arc equal in length to the radius of the circle"; steradian is defined as a "solid angle having its vertex in the centre of a sphere and subtending on the surface of this sphere an area the content of which is equal to the second power of the radius of the sphere."[35] Should we try to overcome this contradiction by interpreting these measurement units – regardless of their evident distinctness – still as the unnamed numbers 1 in the sense of the explicit formulations of older definitions, we would have to accept the consequences of the first alternative.

Neither can one find a way out of this dilemma by regarding the measurement units of dimensional magnitudes as a certain arithmetical unit.[36] Otherwise we would have to admit that some metrical magnitudes, namely dimensional magnitudes with named numerical values, are measured with the help of measurement units of a definite measurement system, while other, dimensionless magnitudes with the unnamed numerical value 1, are measured by means of this arithmetical unit which is independent of every system of measurement units. We should, therefore, concede that some metrical magnitudes are not magnitudes at all, but only the unnamed number 1. Then it might be more appropriate to speak only of dimensional magnitudes and dimensionless numbers. Inasmuch as we wish to keep the division of magnitudes into dimensional and dimensionless, we may seek a way out by adopting the opposite position as well, according to which, not only the number 1 would be included among dimensionless magnitudes, but also all real numbers. If real numbers are viewed as dimensionless metrical magnitudes, it is not necessary to distinguish numbers and magnitudes. Neither can this possible solution be thus regarded as a suitable one.

We certainly do not aim at exhausting all theoretical problems of dimensional analysis, in particular not the question why physical laws are indeed invariant as to the size of the measurement units of magnitudes that are involved in them. Nevertheless, from the aforementioned exposition, we might still arrive at the conclusion that many difficulties would disappear, should we consistently differentiate between the naming and dimension of metrical magnitudes and measurement units.

Naming is an essential sign of magnitudes, distinguishing them from numbers. It refers to empirical variables and reflects the unity of their qualitative and quantitative aspects, as these manifest themselves, especially in their mutual functional relations, which also constitute the objective basis for the formulation of laws. Unlike the more general notion of naming, the concept of dimension is much more special, one might say perhaps, of a more technical nature. This concept is introduced only in connection with a certain system of fundamental and derived magnitudes, or with a pertinent system of measurement units. Primarily, the concept of dimension expresses a theoretically justified selection of fundamental magnitudes, as well as a relation of derivability that exists in the given system of magnitudes between fundamental and derived magnitudes. Secondarily, this concept represents similar relations between their measurement units.

In this respect, every magnitude is thus characterized by a certain name, regardless of whether it is a fundamental or a derived magnitude in the given system, or a magnitude defined in terms of *multiples of powers* of dimensions of fundamental magnitudes, or, finally, whether it is delimited only by a ratio of magnitudes of the same dimension. This holds analogically also for measurement units. Therefore, dimensionless magnitudes, even if their measurement units are numerically mapped by the number 1, always have some name, and this makes them distinct from numbers. The name of the measurement unit of a plane angle – radian – differs from the name of the measurement unit of the level of acoustic pressure – decibel – and it also differs from the name of the measurement unit of the index of refraction, which is not explicitly stated, though their numerical values are expressed in the same way.

Just as unproblematic will be our dealing with the different names of magnitudes of the same dimension, or with the same name of magnitudes of different dimensions, in either case expressed by means of the same or distinct measurement units. This differentiation, as well as the apparent discord between the names of magnitudes and measurement units on one hand, and their dimensions on the other, damage neither the invariance of dimensions of derived magnitudes as to distinct measurement systems nor their relativity with respect to a system of fundamental magnitudes. The concepts of name and dimension, whether they refer to magnitudes or to their measurement units, fulfill *different functions* in the theory of metrical magnitudes which cannot be identified for reasons of simplicity.

3.5. THE CLASSIFICATION OF MAGNITUDES

Occasionally we have mentioned the classification of magnitudes when we talked, for example, about discrete, metrical and non-metrical, physical and extraphysical, fundamental and derived, dimensional and dimensionless, scalar and vector magnitudes, and the like. We might still complement this by a further division of magnitudes, for instance, into extensive and non-extensive, additive and non-additive, primary and secondary. All these classifications, based on different, often very close and sometimes even identical criteria, emphasize the diversified, disjunctive characteristics of magnitudes, always mutually associated in pairs. These characteristics are judged from the most varied viewpoints of content and under the influence of considerably differentiated theoretical positions. Therefore, it could hardly be expected that one might unify and mutually harmonize them to the extent that it would be possible to propose some taxonomy of magnitudes which would be exhaustive as to the relevant problems, logically correct and methodologically appropriate.

For the classification of essential marks of the concept of magnitude from the viewpoint of a general theory of measurement, it will suffice to limit our discussion to both chief classificational levels and their mutual relationship. When the relations between magnitudes are investigated in accordance with the problems of whether and how they represent a unity of qualitative, weakly and strongly quantitative aspects of objectively existing entities, a distinction is being drawn between metrical and non-metrical, extensive and intensive, and additive and non-additive magnitudes. When considering the mutual relations between magnitudes within a definite system of magnitudes, one differentiates between fundamental and derived, primary and secondary, and dimensional and dimensionless magnitudes. Some of these classifications are identical in content and differ only terminologically.

Traditionally, these classifications most frequently start from the difference between *extensive* and *intensive* magnitudes, yet they are differentiated essentially due to the ways of interpreting the content and extension of the concept of an intensive magnitude, depending on the various conceptions of measurement.

Intensive magnitudes are sometimes identified with qualitative aspects and extensive magnitudes with weakly or strongly quantitative aspects. Then among intensive magnitudes one includes purely qualitative properties, for instance, Pilsner, which cannot be regarded as a magnitude; the same holds for weakly quantitative properties admitting the discrimination between degrees, for instance, hardness,[37] and thus non-metrical magnitudes as well.

At another time, the extension of the concept of intensive magnitudes is limited to the concept of a non-metrical magnitude. The existence of a measurement unit is then a classificatory criterion facilitating the decision concerning some magnitude, whether it is in this sense intensive or extensive. For intensive magnitudes, one cannot meaningfully establish any measurement unit; at most, one might establish a unit of ordering. From the viewpoint of a formal theory of measurement, this criterion is expressed by two distinct transformations under which the form of a scale remains invariant (Section 7.2). Intensive magnitudes are then those magnitudes to which one can unambiguously assign numbers up to some monotonic transformation. One can also assign numbers to extensive magnitudes which are unique up to every similarity transformation (i.e., multiplication by a positive constant).[38]

The extension of the concept of intensive magnitudes is even conceived in a narrower way. Although one can stipulate some measurement unit for these magnitudes, one cannot find an empirically meaningful and operationally realizable counterpart of the numerical operation of addition for them. Since in this interpretation, intensive magnitudes constitute a specific subclass of metrical magnitudes, namely non-additive metrical magnitudes, the same meaning is sometimes assigned instead to the so-called non-extensive or in-extensive magnitudes. It also follows from this conception that extensive magnitudes are identified with additive magnitudes and non-extensive magnitudes with the non-additive ones.

In Campbell's view, the discrimination between intensive and extensive magnitudes is connected with the dichotomy of fundamental and derived magnitudes, and this complicates even more the ambiguity that rules here. The lack of clarity and inconsistency of the position itself are responsible for this inorganic combination of two, conceptually distinct levels of classification which rest, in the first case, on the criteria of a measurement unit and an empirical realizability of the numerical operation of addition and, in the second case, on the mutual bonds of derivability in a definite system of basic and derived concepts. On one side, Campbell explicitly declares that no essential and precise difference exists between the fundamental and derived magnitudes and assumes that one might regard as generally valid the fact that, in spite of important exceptions, fundamentally measurable magnitudes are quantities, whereas derivatively measurable magnitudes are qualities.[39] On the other side, however, he extensively analyzes the difference between basic (fundamental) and derived magnitudes, depending upon his conception of fundamental and derived measurement, and tries to find the reasons why it is necessary after all to select a subset of fundamental magnitudes in a set of magnitudes.[40] A bond between both these standpoints is provided by Campbell's view on the function of additivity in the process of measurement; yet even in the exposition of this problem, his conclusions are not entirely consistent.

From the philosophical standpoint, this contradiction can be characterized as a dilemma between the operationalistic approach toward the explication of measuring procedures and a naively materialistic attitude toward the question under what empirical conditions could one rightfully assert that the assigned numerical expressions grasp the quantitative aspects of the measured objects.

In the operationalist conception of measurement, there is no essential difference between basic and derived magnitudes. Of course, this doctrine ignores objective presuppositions and theoretical reasons which authorize us to regard some magnitudes in the given system as basic and others as derived. Although this relativization can be operationally substantiated by the possibility of measuring many magnitudes fundamentally as well as derivatively, it is nonetheless incompatible with the theoretical construction of physics in which this classification is still applied. Such relativization is also in conflict with the fact that the discernment of fundamental and derived magnitudes has, at any rate, a principal significance for the construction of coherent systems of measurement units. On the other hand, the emphasis, in the operationalist position, on the empirical interpretation of the numerical operation of addition – which reduces the arbitrary nature of the numerical assignment in the process of measurement to the choice of the unit of measurement – is undoubtedly an expression of its effort to give an objective justification for measurability of one or the other property existing in reality, and thus to regard natural law-like regularities as prior to their numerical and theoretical representations.

The difficulties encountered in the differentiation of the various kinds of magnitudes have, however, their objective reasons too. In concrete cases, mutual relations between distinct kinds of magnitudes are not always unambiguous. Obviously, many discrepancies also reflect the different theoretical viewpoints concerning the fundamental question of the extension which one might actually assign to the concept of measurement. No wonder, therefore, that among intensive magnitudes, also interpreted as magnitudes expressing physical states,[41] one finds, not only hardness or the intelligence quotient but also solubility, density, temperature, electric potential, intensity of a magnetic field, etc., and that extensive magnitudes are sometimes distinguished from and at other times identified with, additive magnitudes in accordance with the interpretation of the additivity condition (Section 8.2).

According to our view, which is based primarily on the differentiation of the qualitative and weakly and strongly quantitative aspects of objectively existing objects, as well as on the distinction between magnitudes and numbers, magnitudes can, in general, be put into various subclasses by means of two classificatory criteria.

In accordance with our ability to establish measurement units or units of ordering for magnitudes (which are objectively conditioned by the fact that some of them reflect only the ordering of degrees, while others reflect the ordering of their size, too), as previously mentioned, we distinguish *non-metrical* and *metrical* magnitudes on the first classificational level.

According to the criterion of empirical additivity, we then differentiate the following subclasses of metrical magnitudes on a further classificational level.

Metrical magnitudes which satisfy the condition of empirical additivity without any restriction, for example, length, shall be regarded by us as *strictly extensive*.

We shall call *extensive* those metrical magnitudes which are additive only under certain conditions, for instance, volumes of materials under the assumption that they have the same chemical composition.

Metrical magnitudes which might be regarded only intermediately as additive, and then only under certain theoretical presuppositions, e.g., temperature, shall be called *quasi-extensive*.

In this classification we have consciously stayed away from the ambiguous term 'intensive', which, in one of its meanings, agrees with the concept of non-metrical magnitudes, yet, in another, with the notion of quasi-extensive magnitudes. For similar reasons, we have not considered it necessary in our trichotomic classification (which unifies, in a certain way, the division of magnitudes into extensive and non-extensive, additive and non-additive), to express explicitly in the names of particular subclasses also their common classificatory basis, namely the criterion of empirical additivity.

When talking about fundamental and derived magnitudes, we must be well aware that this classification of *magnitudes* cannot be understood in the same sense as the classification of the various *kinds of measurement* or *measuring procedures* (Section 6.2), where we likewise discriminate between fundamentally and derivatively measured or measurable magnitudes or, more precisely, between fundamental and derived measurement and between fundamental and derived measuring procedures, in spite of the fact that we might be easily tempted to proceed in this direction, not only due to the terminological similarity, but also due to mutual connections.

What are the conditions on the basis of which one may distinguish the classes of *fundamental* and *derived* magnitudes? Originally, Campbell gave the following criteria for these classifications:[42]

(i) *The manner of their measurability*. Fundamental magnitudes are measured fundamentally, i.e., independently of the foregoing measurements of the same or some other magnitude, by some fundamental measurement procedure, i.e., by means of some direct, immediate way of measuring; derived magnitudes are measured according to the measurement of other magnitudes,

(ii) *The satisfaction of the laws of addition*. For fundamental magnitudes it is possible to find an empirical counterpart of the numerical operation of addition but not, however, for derived ones,

(iii) *The relation to numerical laws*. Derived magnitudes are measured in terms of constants of numerical laws; fundamental magnitudes are independent upon numerical laws.

With regard to these three criteria, we assign the greatest significance, though with certain reservations, to the *first two* which are essentially equivalent in meaning.

On the basis of these two criteria, Campbell then presents a "complete list" of fundamental (physical) magnitudes, divided into two groups. Into the first group he puts length, time, weight, and electrical resistance, while into the second group he puts 17 other metrical magnitudes: angle, area, volume, energy, mass, moment of inertia, momentum, electrical capacity, electrical current, potential difference, electrical conductance, magnetic induction flux, etc. All these fundamental magnitudes can fulfill the condition of additivity and may be measured by means of a fundamental process of measurement, yet only the first four are measured in this way.

We do not regard these two criteria as appropriate or sufficient for this classification of magnitudes. The difference between fundamental and derived measurement refers, in principle, to kinds of measurement or to measuring procedures and not, however, to a division of magnitudes within the system of (physical) magnitudes. Yet there exists no one-to-one correlation between the fundamental and fundamentally measurable magnitudes, nor between the derived and derivatively measurable magnitudes. There exist namely fundamental magnitudes which are fundamentally measurable, e.g., length, yet there are also fundamental magnitudes, e.g., temperature, which are not measurable in this manner.[43] The requirement of additivity is a distinguishing criterion in the classification of metrical magnitudes, yet it does not refer to the differentiation of fundamental and derived magnitudes. Otherwise we would indeed have to concede that every fundamental magnitude must always be additive and, thus, that these magnitudes could only be strictly extensive ones. Then we could not regard as a fundamental magnitude, for example, temperature, which is generally viewed as an intensive, i.e., quasi-extensive magnitude.

These reservations concerning Campbell's doctrine are not weakened by the fact that later, in connection with the exposition of dimensional analysis, he introduces the term *basic magnitude* besides the term *fundamental magnitude*,[44] in order to be able to distinguish between the so-called practical basic magnitudes (which are measured fundamentally) and the so-called theoretical basic magnitudes (which are held to be basic in the system of [physical] magnitudes). Nevertheless, no one-to-one correspondence exists between practical and theoretical basic magnitudes. For instance, although on a theoretical level mass corresponds to weight, which is a practical basic magnitude, no practical basic magnitude corresponds to temperature, which is to be regarded as a theoretical basic magnitude. In another connection,[45] Campbell even declares that this magnitude, in the strictest possible sense of the term, is actually no magnitude at all. It does not satisfy, namely, the fundamental requirement which is imposed by Campbell on the metrical, fundamentally measurable magnitudes – the requirement of additivity. This contradiction is also manifested if we apply this conception to derived magnitudes. Thus, for example, volume is an additive magnitude and can, therefore, be measured fundamentally. Hence, it meets the sufficient conditions imposed on every practical basic magnitude. Yet no theoretical basic magnitude in the foregoing sense can be assigned to this derived magnitude.

Of course, Campbell is well aware of all these inconsistencies, yet even his later attempt to explicate the concept of a basic magnitude[46] remains unsuccessful. Now he lays out the following criteria for the inclusion of a magnitude in the class of basic magnitudes:

(i) a concrete way of measuring a given magnitude – basic magnitudes are measured independently of the measurement of other magnitudes, i.e., fundamentally;
(ii) the relation to the system of measurement units – basic measurement units are conventionally assigned to basic magnitudes;
(iii) the theoretical significance that is assigned to individual magnitudes in accordance with the hierarchical arrangement of physical laws – basic magnitudes occur in the class of primary physical laws or in other classes linked up with this primary class, unless they had already occurred as derived magnitudes.

These criteria, which evidently are not disjunctive, do not, however, delimit the same class of basic magnitudes, into which Campbell now includes number, mass, length, time, electrical resistance, and angle. Considering these magnitudes, only length, mass, and time fulfill these criteria and not, however, number, which is not a magnitude, nor electrical resistance or angle, which do not satisfy the second and third conditions.

However, how can one explain on the basis of this doctrine that temperature too belongs among the basic magnitudes of the current system of physical magnitudes and, yet, temperature does not satisfy the first criterion? By virtue of accepting the criterion of fundamental measurability for the classification of magnitudes – a criterion which implies that every basic magnitude must satisfy the condition of additivity – Campbell obstructs his own path leading towards the solution of this particular question and nothing else remains for him but to claim that, contrary to the theory, temperature is not a basic magnitude. Since at the same time he acknowledges, to the contrary, that we could also fundamentally measure magnitudes which ordinarily we do not consider basic, as well as that not all magnitudes which are "basic in theory" are also "basic in measurement", he cannot adequately explain the mutual relationship between basic and derived magnitudes on a general level. Campbell attempts to paper over this conflict between the theory and practice of measurement by relativizing this classification of magnitudes and declaring without further argument that magnitudes which are basic, in accordance with the above-mentioned criteria, are simply, by coincidence, very important to us.

If therefore we have already admitted that the principal difference between basic and derived magnitudes hinges upon the issue that basic magnitudes must, of necessity, satisfy the condition of fundamental measurability and that, consequently, we must always find for them some empirically meaningful and operationally realizable counterpart of the operation of addition, then, as has been stated, temperature cannot be included. One cannot inconsistently dodge this conclusion (which inevitably follows from the primary classificatory criterion of Campbell's conception) by distinguishing two different concepts of temperature. This position is characteristic of many authors. Thus, for example, Smart[47] assumes that temperature, so far as it is measured with the help of the so-called thermodynamic scale, is to be regarded as a theoretical magnitude; yet he does not mention explicitly whether in this case it is or must be an additive magnitude, too. If it is measured by means of the so-called gas scale, Smart holds it to be a derived magnitude which, beyond any doubt, need not be additive and, in fact, is not additive in the given case.

If in the current system of physical magnitudes, temperature is considered a basic magnitude, this decision does not really depend on the question of whether it is measured fundamentally, derivatively, or associatively. However, this does not depend on the satisfaction or non-satisfaction of the condition of additivity, required by Campbell, either. If we include temperature among basic magnitudes, we do so because – for objective, theoretical, and practical reasons, conditioned by the historical development of physical knowledge – a conclusion has been reached that it is quite necessary to assign to it the function of a basic, primitive notion in this system of metrical magnitudes.

The analogy between basic and derived magnitudes in the system of physical magnitudes on one hand, and basic and derived concepts in some axiomatic system on the other, has, of course, only a limited validity. In an axiomatic system (consider, for example, an axiomatization of propositional logic), various sentence-producing functors may be selected as primitive notions, e.g., only incompatibility, or implication and negation, or implication, negation, conjunction, alternation and equivalence, while the individual equivalent axiomatic systems are in no way essentially different from the theoretical and practical viewpoints. In the system of physical magnitudes, however, the possibility of choice is restricted because of objective causes. Nevertheless, this restriction cannot be interpreted as if there were an absolute difference between basic and derived magnitudes. Within the system of physical magnitudes one might, of course, accept electrical charge instead of electrical current as a basic magnitude of the subsystem of electromagnetic magnitudes. Yet, without the system of mechanical magnitudes, one could scarcely construct a consistent system of physical magnitudes which would be in harmony with the theory and which would also meet the requirements of practice. And, if the system is not based on length, then, apparently, we could not speak at all about a physical system.

The choice of basic mechanical magnitudes is not an expression of some convention and it does not depend upon the manner of a theoretical introduction or upon an empirical measurement of these magnitudes. Nevertheless, this choice is a consequence of the fact that every real object exists in time and space. If we take into account only operational procedures, by means of which they can be measured, and if we ignore the priority of objectively existing quantitative aspects of the measured objects, we may arrive at an entirely opposite position [48] and assign to the mutual relationship between basic and derived magnitudes – in contradistinction to the analogical relationship between primitive and defined concepts – an even greater degree of arbitrariness.

In the system of metrical magnitudes we might also consider various other divisions. If, for instance, we investigate mutual relations between basic and derived magnitudes from the standpoint of dimensionality, basic magnitudes are dimensionally *simple*, while derived magnitudes are dimensionally *complex*. This also refers to derived magnitudes when they are proportional magnitudes, i.e., dimensionless derived magnitudes. We may further divide dimensional derived magnitudes into *one-dimensional* if they are defined with the help of dimensions of only one basic magnitude, and into *many dimensional* if they are defined by virtue of dimensions of at least two different basic magnitudes. For example, volume or area belong into the class of one-dimensional derived magnitudes, whereas a majority of other derived magnitudes, for example, force, velocity, and the like, fall into the class of many-dimensional magnitudes.

However, these classifications, to which further divisions might be added as well, no longer essentially extend our knowledge concerning the nature and function of magnitudes from the point of view of a general theory of measurement.

4. SCALES

The differentiation of metrical and non-metrical magnitudes on the empirico-mathematical level of the conceptualization of the object of measurement must necessarily appear in an explication of the result of measurement, too. Different numerical relational structures will correspond to the degrees of non-metrical magnitudes more than those which correspond to the sizes of metrical magnitudes. In either case, however, these structures constitute a certain progression of numerical values which are comprehensively designated by the term *scale*.

The concept of a scale is a further basic concept to which great significance is assigned, particularly in the theory of measurement in the widest sense of the term. By scale one understands not only the results of measurement procedures applied to a certain empirical relational system but also a tool of measurement which is used in measuring some magnitude, or, if need be, only the scale of a certain measuring apparatus. This semantic ambiguity further increases when the concept of a scale is also interpreted in the same or similar sense as the concept of measurement.

4.1. THE CONCEPT OF A SCALE

The concept of a scale, which is ordinarily given the same meaning as a *standard* or a *gauge*, is specified in the theory of measurement in various ways. Generally, one speaks either about *scales of measurement* or *measuring scales*, more specifically, about scales of measurement of a certain magnitude, for instance, about time scales, gas scales for measuring temperature, and the like. When one talks about Guttman's, Thurston's, Likert's, or Edwards' scale[1] in the theory of extraphysical measurement, for example, the concept of a scale is identified with the concept of scaling, namely, with the procedure or technique used by the respective author. Sometimes the concept of a scale of measurement is identified with the concept of magnitude; for example, hardness is regarded as an ordinal scale, while the gas scale of temperature is declared to be a derived magnitude.[2] In the formal theory of measurement, the concept of a scale is interpreted in the same sense as the concept of measurement.[3] Scale is here considered an ordered triple

$$\langle \mathscr{E}, \mathscr{N}, \Phi \rangle,$$

where $\mathscr{E}$ signifies an empirical relational system, $\mathscr{N}$ a numerical relational system, and Φ an isomorphic or homomorphic mapping from $\mathscr{E}$ to $\mathscr{N}$.

In order to explain the function of scales in the theory of measurement, we must first clarify the difference between so-called *metrical* and *conceptual* scales.[4]

In geometry, where the measurement procedure was practically born, the scale was a certain standard, gauge, or means of measurement by means of which objects of different length had been compared. This comprehension of scales was also applied later, when the practice of geometrical measurement had been extended into various physical domains. A scale was simply conceived as a certain measuring instrument or, more precisely, as a specific component of such an instrument. However, as soon as measurement had been extended into social scientific areas and investigators also began to measure such non-metrical magnitudes as, for example, intelligence, sensations, opinion, and attitudes, for which, currently, one could hardly construe some measuring instrument, it was not possible to interpret the concept of a scale in the sense of a scale of a measuring instrument. Only now has the concept of a scale acquired a more versatile function from both theoretical and methodological viewpoints.

The distinction between the material form of a scale and its conceptual appearance is quite obvious if we consider, for example, the difference between the dial of a clock and the scale of time measurement. Insofar as we have some suitable means of measurement, some standard, graduated or calibrated object, with respect to which something is objectively compared and arranged, to every pair

<magnitude, measuring instrument>,

for example, <length, yardstick>, <temperature, thermometer>, <pressure, barometer>, <density, pycnometer>, we can assign two distinct scales – conceptual and material. This differentiation, which does not deny their mutual connections conditioned by the bonds between the empirical and theoretical aspect of measurement, refers, above all, to the scales of metrical magnitudes.

Since in the Czech language, besides the term *škála*, there is also the term *stupnice*, we consider it convenient to introduce the following terminological convention: by the term *škála*, we shall signify conceptual scales, while by the term *stupnice* we shall signify material scales.[5]

A *conceptual scale*, in short a *scale*, is characterized by a certain ordered interval of numerical values, the so-called *scale values*, which can be theoretically assigned to the measured magnitudes, depending on their qualitative and quantitative aspects and on their conceptual fixation within a definite theory. The notion of a conceptual scale cannot be identified with the concept of numerical scale, i.e., with a certain ordered interval of numbers, for the scale values might be the same for different magnitudes, even if the objective properties of the measured magnitudes and their interpretation within the theory conspicuously differ. This is obviously due to the fact that the domain of variability of scale values is homogeneously restricted to the set of ordinal or cardinal numbers or to some of its subsets, and is distinguished only by virtue of the scope of a given numerical interval, while the domain of variability of the measured magnitudes is qualitatively very heterogeneous.

The *form* of a conceptual scale, indicating the interpreted distribution of scale values, is delimited by the character of these numerical values, the origin of the scale, the so-called scale zero which limits the ordered interval of scale values usually on the left-hand side, and by the arbitrary nature of the distance between any two immediately adjacent scale values, contingent on the measurement unit. The limitation of this ordered interval on the right-hand side, which is to be assumed, at least for practical reasons, is not essential for stating the form of scales.

A *material scale*, in short a *gauge*, is determined by an ordered set of marks on the measuring instrument, and in most cases by a set of numerals the reading of which enables us to assign numerical values to the size of the measured magnitudes. The scope of these marks and their arrangement, especially their spacing, depends on the relationship between the material scale and the respective conceptual scale. It is also contingent upon the constructional possibilities of instrument technology, their concrete materialization and use in the process of measurement, as well as upon the requirements of the desirable or attainable degree of exactness of measurement.

From the given explications of the concepts of scale and gauge there follows, in the first place, a very essential distinction which we should not neglect: Material scales are characterized by *numerals*, whereas in the case of conceptual scales we deal only with *numbers* – either ordinal or cardinal. The numerical scale values of conceptual scales facilitate the specification of the ordering or size of non-metrical and metrical magnitudes. Numerals alone cannot fulfill this function because we can also use them, regardless of either of these possible interpretations, only as conventionally chosen signs which we by no means assume are names of ordinal or cardinal numbers. Scale values cannot be linguistic expressions, only extralinguistic entities – numbers. Numerals are, at most, the names of scale values.

Another problem which follows from the differentiation of scales and gauges is the question of their mutual relationship. Is this relation unambiguous, one-to-one, or is it not? Only a consistent advocate of the operationalistic doctrine of measurement will defend the view that exactly one gauge will correspond to a definite scale, and vice versa. This conception is unsubstantiated. A gauge with an identically structured set of numerals may, after all, correspond qualitatively to very heterogeneous scales. The very gauge of a measuring instrument, due to the ambiguity of its meaning, is practically unusable unless it is interpreted with regard to the weakly or strongly quantitative aspects of a measured magnitude of a certain kind. Imagine only two different measuring instruments with the same gauges. If we do not know the name or the dimension of the measured magnitudes and their measurement units, obviously we cannot say to what the numeral signs on the gauges of measuring instrument refer. If we take into account the most varied measurement procedures, tied up with the use of diverse measuring instruments, it is further evident that different gauges will correspond to one and the same scale. Some of them will serve for obtaining the same scale values; others, in turn, will be constructed precisely for the sake of furnishing scale values which lie in very distinct numerical intervals and are accessible only under the application of different measurement procedures and measuring instruments. A certain magnitude might be measured by means of distinct gauges, more precisely, the gauges of the measuring instruments, and yet one and the same magnitude, characterized by a single scale, might be at stake here.

A one-to-one relation between scales and gauges can be obtained only when we move from individual scales and gauges to classes of scales – to scale types (Section 7.1). However, by virtue of this abstraction, the qualitative difference between individual scales and gauges gets lost. This abstraction is, therefore, acceptable only in determining structural properties of various gauges and scales with respect to the form of the respective scale type.

Since scale values of scales may be only numbers expressing either ordering or size of the measured magnitudes, we shall distinguish in accord with our previous classification of magnitudes, two kinds of scales: *non-metrical* and *metrical*. We shall also call non-metrical scales *topological* scales.

The differentiation of scales and gauges can be considered, strictly speaking, only for metrical magnitudes, since only these magnitudes are measured by us with the help of different measuring instruments.

4.2. THE ORIGIN OF A SCALE

For the clarification of the form of a scale we must further analyze the concept of the *origin of a scale* or the so-called *scale zero*. While discussing these problems, we clearly should investigate first of all the mutual relations between the scale zero and other species of zeros. At the same time we should assess the connection between the concept of zero and a whole range of related notions, above all, the concept of empty class. Since the scale zero is usually identified with the numerical zero (with the number 0), it will suffice to seek answers only to two questions. How could this specific scale value be empirically interpreted? Can one arrive in the process of measurement at scale values smaller than zero? Both these questions might be formulated summarily as follows: Can we find in the objective reality a certain empirical counterpart of the numerical zero or of negative numbers which would have an empirically significant and non-trivial interpretation?

While the objectively existing objects are always qualitatively and quantitatively determined in a certain manner, various species of numbers are, in spite of their real foundation, essentially idealizations and one cannot find for them corresponding empirical counterparts with *precisely* the same structural properties. If we accept the validity of this presupposition, it will appear obvious to us that objectively there cannot exist any empirical property which would in itself have some zero, i.e., no size. Even less could there exist some property that would have a size which would be even smaller than the zero size. Objects existing in reality do not have "zero length" or "zero weight", nor have they in this interpretation "zero temperature", for they are – with regard to a given property, if it, of course, belongs to them – always *some* and never *none*. For this reason we should look for a meaningful interpretation of the number 0, if it appears as a scale value, either in the process of measurement or in relation to the properties of the respective empirical relational system.

In the first, rather trivial case we might simply say that under an empirical counterpart of the numerical zero one ought to understand an absence of a certain property in the given empirical system. This may be, for example, ascertained in the process of measurement by observing that the pointer of the measuring instrument is not deviating from its initial state. This interpretation holds, of course, under the assumption that the measuring instrument is not damaged and that it is sensitive enough to react to the occurrence of a property in question.

In the second, non-trivial case, we must realize first that the scale zero cannot be generally interpreted in the same sense as the number 0, and that we must refer it to the pertinent empirical relational system. This also holds for negative scale values which should not be identified with negative real numbers, since otherwise we would have to admit the existence of negative magnitudes in an ontological sense. At any rate, we must differentiate between those situations when the numerical values corresponding to different sizes of metrical magnitudes refer to other numerical values which are assigned to these magnitudes within a given relational system, and those situations when this is not the case. Ignoring either of these possibilities easily leads to inconsistent conclusions. Thus, for example, although Campbell [6] rejects the viewpoint that there exist bodies having "zero length" or "zero volume", he nevertheless admits the existence of bodies with "zero weight". Likewise, he does not acknowledge the existence of negative lengths or negative volumes, yet he admits the existence of negative distances or negative time intervals.

To avoid this doctrine, we accept Menger's [7] notion of an *oriented class of magnitudes* which assumes the possibility of constructing two positive classes of metrical magnitudes that depend only upon the measurement unit but do not comprise among their elements the zero element, the so-called *nil* (null). Between these two classes there exists the following relation: To each element of one of them there corresponds some element of the other class and vice versa, and the element of one class is the converse of the correlated element of the other class.

Both these contrary classes, specified, for example, as a class of objects heavier than air and a class of objects lighter than air, which are as such positive, may be united into a *complete* class. Only for this class is it possible to give a meaningful empirical interpretation of the zero element. This interpretation is provided by an intersection of two mutually corresponding contrary positive classes; for example, for weighing it is the equilibrium state on the lever balance. In ascertaining distances, both classes are characterized by various distances to the left or to the right, westerly or easterly from a certain place, for example, from the place which lies on the Greenwich Meridian. In the case of time intervals, both these contrary classes are represented by the classes of past and future events or by the classes of temporal instants before and after a given temporal instant. The zero point might then be understood either as the contemporary temporal instant or rather as a point to which individual temporal instants refer.

We might therefore regard the scale zero either as the point of equilibrium or as the point of reference. The choice of the interpretation will depend on the objective properties of the measured magnitude, as well as on the specific theory of measuring this magnitude. If we can identify the scale zero as the smallest scale value that is empirically accessible and theoretically presupposed, with the numerical zero (and that is always convenient for reasons of calculation), then we can assign positive numbers to elements of one class and negative numbers to elements of the other, oppositely oriented class. One cannot infer from this, however, that magnitudes with negative numerical values objectively exist too.

On the basis of this conception one can also clarify the difference between an *arbitrary* scale zero and a *natural, absolute* scale zero, which is often interpreted also as the difference between those progressions of sizes of magnitudes that objectively have some minimum (e.g., temperature) and those progressions that do not have it (e.g., time).

In talking about an arbitrary zero, an arbitrary origin of a scale, it is assumed that on the scale, conceived either geometrically as a progression of points or arithmetically as a progression of numbers, there exists some point or number conventionally chosen as the most suitable origin of the scale. Scale values, ordered in a numerical series, are then oriented from their beginning to the right and to the left. For scales with a natural (absolute) origin this scale value is located at one end of this progression. The scale is oriented only in one direction, to the right from its origin.

The differentiation between an absolute and an arbitrary origin as points of reference of metrical scales is usually exemplified by the Celsius, Réaumur, and Fahrenheit scales of temperature on the one hand, and the Kelvin and Rankine scales on the other. The choice of different numerical values for the origin of these scales – 0°C, 0°R, 32°F and –273, 16°C – is conditioned by various factors: The construction of thermometers and their practical use, gradual constitution of our ideas concerning the nature of temperature, the requirements of calculation and; in particular, the consequences of the theory of thermodynamics. Regarding all these circumstances, can we draw a sharp boundary between an arbitrarily selected origin, for example, of Celsius' scale, and the absolute (natural) origin of Kelvin's scale? Without any doubt, a progression of possible temperatures has its minimum, if we start from the presuppositions of thermodynamics. Likewise, there is no doubt that, if we take Dalton's caloric theory and also accept with it his "logarithmic scale" which meets all objectives of measuring temperature and fully satisfies the requirements of calculations, we would never be compelled to assume that a progression of thermal states is limited downwards.[8] It is quite possible that, under these theoretical presuppositions, the idea of introducing the concept of absolute zero would have never arisen in our mind. The reasons for preferring the Kelvin or Rankine scales over the Celsius, Fahrenheit, or Réaumur scales, as well as for neglecting Dalton's "logarithmic scale" are, above all, of a theoretical nature.

However, if we take into account the purely practical usage of temperature scales in the various uses of thermometers, the arbitrary origin of the Celsius scale will appear to us much more natural. Consider, for example, the scale for measuring the temperature of the human body, the values of which lie, say, from 35° up to 42°C. For the sake of measurement of this kind, the value 35°C may be held to be the origin of the scale. We might even quite arbitrarily re-scale its values so that we would interpret this origin as the scale zero of this temperature scale. If we compare the origin of the Fahrenheit scale with the beginnings of the Celsius and Réaumur scales, then the scale zero 32°F will appear to us much more arbitrary than the scale zero 0°C (or 0°R) which corresponds with the beginning of the numerical series.

The differentiation between two species of *scale zeros*, which is obviously contingent upon our ideas concerning the nature and function of the origins and upon a strict dichotomy between what is regarded as arbitrary and what is regarded as natural, leads toward the following dilemma. For scales with an arbitrary zero, for example, the Celsius, we must assume a meaningful empirical interpretation of negative real numbers, whereas for scales with an absolute zero, for example, the Kelvin, we cannot presuppose it. If we consider temperature, in one case, as a negative magnitude, and in another case, as not, we would have to distinguish two different magnitudes of the same name yet with clearly distinct properties.

Undoubtedly, this consequence is unacceptable. As far as we regard temperature as a uniform metrical magnitude, theoretically fixed in the current system of physics, the distinction between different temperature scales can be explained by the fact that, not only different origins, but also different measurement units are stipulated for them. If they have the same origins, the law of invariance as to the change of the unit holds for the scale values of temperature quite generally; if they do not have the same origins, its validity depends on an additive constant which facilitates the conversion of these origins as well. In this sense, then, temperature does not differ at all from other metrical magnitudes, for example, from length, which can be measured by means of measurement units of various measurement systems. Further, it is to be taken into account that the form of a scale cannot determine objectively existing qualitative, weakly and strongly quantitative aspects of a magnitude. Finally, one cannot yet infer the real existence of negative magnitudes from the mathematical existence of negative numbers.

The terms *natural zero* and *arbitrary zero* are not happily chosen. They are sometimes analogically transferred to measurement units and, thus, besides *conventional* units, one speaks about *natural* measurement units.[9] These terms may easily seduce us to illegitimate conclusions, such as those which we encounter especially in the discussions concerning scales of temperature. Further consequences are then derived from the former, purporting to justify measurability of non-metrical magnitudes in the narrow sense of the word, although no measurement unit can be established for them. For this reason, we do not regard the distinction between a *natural* and an *arbitrary* origin of scales as a suitable terminological convention.

The stipulation of the origin of scales, whether they are oriented one-directionally or two-directionally, is a result of ordering some class of magnitudes and, hence, the corresponding scale values too. An *arbitrary* zero as well as the *absolute* zero are theoretical constructs that are neither absolutely arbitrary nor absolutely natural. The difference between them may be expressed only by a greater or lesser degree of arbitrariness, conventionality: *The absolute zero has a less conventional character than the arbitrary zero*. The choice of the origin, always objectively determined in some way, can therefore be more or less empirically justified, more or less suitable with respect to the manner of calculation, and more or less plausible with regard to approved theories.

4.3. DISTANCE

The concept of distance, or the so-called distance function, which plays an important role in defining the form of scales chiefly in the theories of extraphysical measurement, is not interpreted in a satisfactory way either.

On one hand, the concept of distance is used synonymously with the concepts of difference or interval, while on the other hand, it is brought into connection with the concepts of numerical operation of addition and its empirical counterpart. However, in either case it is not evident which of these concepts designates the relation between scale values and which is the one expressing a certain empirical relation.

Thus, for example, Torgerson[10] at one time regards ordering *difference* and origin as characteristic features of numbers, yet at another time, ordering, *distance* and origin. The differences between real numbers are interpreted as the distance between quantities, but a "temperature distance" is defined as the amount of difference in temperature between any pair of quantities. Coombs[11] rather confuses the concepts of distance and interval when he talks about a distance function that assigns some real number to all pairs of elements in some ordered set and is specified as a distance or interval between objects or classes of objects.

Pfanzagl[12] introduces a special concept of *metric connection* and strives to demonstrate that between the concepts of distance and metric connection there exists a relation of mutual definability. If the concept of metric connection is introduced and defined by the conditions of existence, monotonicity, continuity, and bisymmetry, the concept of distance might be defined or specified by the conditions of ordering, monotonicity, continuity, replaceability, and vice versa. In order to characterize this analogy, it will suffice to introduce from the axioms of the theory of metrical connection, which are structurally in agreement with the axioms of the theory of mean values,[13] the axiom of bisymmetry

$$(a \circ b) \circ (c \circ d) = (a \circ c) \circ (b \circ d), \tag{4.3–1}$$

and from the axioms of the concept of distance the corresponding axiom of replaceability

$$\text{from } a\,b \gtreqless c\,d \text{ there follows } a\,c \gtreqless b\,d, \tag{4.3–2}$$

formulated for the pairs $a\ b \in M$, where $a\ b \neq b\ a$. The concept of metric connection is taken for the generalization of the concept of additive operation. Every additive operation is a metric connection just as is the operation of constructing the midpoint, the operation of bisection, utilized, for example, in psychophysical measurement, in determining that stimulus which lies exactly in the middle between two given stimuli, or in measuring partial beliefs, values, desires, utilities, and the like, by means of Ramsey's method which is based on the concept of the degree of belief $p = 1/2$ in an ethically neutral statement.[14] The operation of constructing the midpoint must, in addition, fulfill the condition of commutativity and reflexivity. With respect to the mutual definability of the concepts of metric connection and distance and with the help of the concept of distance, one might also define the concept of a midpoint and the additive operation. In the first case, the issue is the connection between elements of an abstract set M, interpreted as a set of objects which have some property in a certain intensity. In the second case, what is involved is the connection between equivalence classes of pairs of elements of this set.

Before we shall attempt to expose the reasons which led to the introduction of the concept of distance into the theory of scales, we must mention that this concept, as it has been defined on the basis of the concept of a segment, for example, by Hölder[15] (not only Pfanzagl, but also Suppes and many other authors refer to his axiomatization) has its primary application in theoretical mathematics. The concept of distance is generally understood as a system of relations between ordered n-tuples of real numbers $(x_1, x_2, \ldots, x_n)$ and $(y_1, y_2, \ldots, y_n)$. In the case of a Euclidean distance between two points, it is then defined by the expression

$$\sqrt{\sum_{i=1}^{n} |x_i - y_i|^2}. \tag{4.3–3}$$

The concept of distance is also defined in set-theoretical topology. If there exists a function $d(x, x')$ on the product $X \times X$ with values in the set of non-negative real numbers (where X is a non-empty set with elements $x, x', \ldots$), satisfying the following three conditions:

$$d(x, x') = 0 \quad \text{exactly when } x = x' \tag{4.3–4}$$

for all

$$x, x' \in X : d(x, x'') = d(x', x) \tag{4.3–5}$$

for all

$$x, x', x'' \in X : d(x, x') \leqq d(x, x'') + d(x'', x'), \tag{4.3–6}$$

we say of this function that it is a *metric on X*. The set X together with the *metric on X* is then called a *metric space*.

In both these cases the concept of distance is quite undoubtedly an idealized mathematical concept. The concept of distance $d(x, x')$ has primarily an adequate empirical interpretation and a significant application only as far as we understand it as a geometrical concept. Any further extension – unless this concept is appropriately modified in its meaning – is already questionable to a considerable degree. At any rate, this also holds for the concepts of interval and difference, at least in their technical usage. These concepts are arithmetical and not empirical. If they are used outside the area of mathematics with distinct meanings, they should at least be redefined for those applications specified for pertinent domains. Just as clearly, it should have been said which of these concepts are empirical and which are numerical; to express it differently, which concept refers to the properties of the measured magnitudes and which concept refers to the scale values that are assigned to them.

The inclusion of the geometrical concept of distance in the conceptual network of the theory of scales, unfortunately only on the basis of a fairly superficial analogy, might be first of all explained by the confusion between scales and gauges. It is well known that the reading on the registering equipment of measuring instruments may be easily realized by "measuring" the distance between two material points, for example, equal distances between the marks on the thermometer tube. This is, of course, a result of observing the spatial relation of distance, measurable as the length of the corresponding material segment. If we use a gauge in this sense, we can – as it is sometimes metaphorically expressed – "reduce every kind of measurement to the measurement of lengths". Yet this possibility definitely does not permit the inference to the effect that the concept of distance (or, more precisely, the concept of length distance) is empirically interpretable in other cases, for example, if we compare various objects according to their temperature, density, and the like.

It is sometimes claimed that it is admissible to draw, without any further specification, relevant conclusions from the properties, relations, and operations with numbers as to the corresponding properties, relations and operations with the measured magnitudes. We have already hinted at the illegitimacy of this position. Let us admit, however, that for different magnitudes we have at our disposal some empirical notion of distance, corresponding to the mathematical concept of distance or, better to say, to a difference or an interval between real numbers. Can we even register empirically every theoretically conceivable distance between numbers? It is quite evident that *equal* distances between real numbers will sometimes be empirically significant and measurable, yet, at another time, they will not. Consider only equal distances between two pairs of numbers, for example, between the numbers 9 and 10, and between the numbers 1 000 000 000 000 009 and 1 000 000 000 000 010. Will these distances be empirically comprehensible in both cases in the same way? This will also hold analogically for *various* distances between "very small" and "very large" numbers. Their difference will be empirically discernible, measurable, and meaningful only within certain limits, depending on the nature of the measured magnitude as well as on the practical needs and aims that we pursue at the same time. In looking for an empirical counterpart of the mathematical concept of distance, we have to establish the boundaries within which we are able to discriminate equal and different distances and, at the same time, stipulate when the differences between numerical distances are empirically relevant and when are they not.

A further, apparently the most weighty, reason is undoubtedly the effort to extend the concept of measurement, in the narrowest sense of the word, to non-metrical magnitudes. Because there are no measurement units for these magnitudes, still less can they satisfy the condition of additivity. If we cannot find an empirical interpretation even for such a simple arithmetical operation as the operation of addition, we can hardly speak of a homomorphism (or isomorphism) between the numerical relational system and the empirical relational system. To satisfy this fundamental condition of measurability, one has to look for some suitable substitute for the operation of addition, as well as for the measurement unit. And it is exactly this purpose which should be served by the concept of distance (distance function) or the concept of metric connection.

In introducing the concept of distance one might refer, at least *per analogiam*, on one hand, to Hölder's theory of extensive magnitudes in which, of course, *distance* is always a metrical magnitude satisfying the condition of additivity and, on the other hand, to Carnap's conception of the construction of physical concepts, which is linked to Russell's distinctive theory of magnitudes and quantities, in which, in turn, the differentiation between distances and "divisibilities" is brought into a relationship with the difference between additive and non-additive properties of magnitudes. Carnap[16] distinguishes two different subclasses of distances: distances that are metrical magnitudes (e.g., length distances) and distances that cannot be regarded as metrical magnitudes (e.g., temperature distances).

It is to be noted, as against this position, which is apparently intended as an attempt to clarify the difficulties encountered in the traditional exposition of measuring temperature, that Russell[17] himself held another viewpoint. For him, distance is a metrical magnitude, just like the difference between magnitudes. However, distance is oriented, whereas difference is not. The contention that temperature distances are not metrical magnitudes also conflicts with the conception of interval scales (Section 7.1). This scale type is precisely characterized by the condition that arithmetical operations, which in this case cannot be performed on scale values themselves, are exactly applicable only to distances (differences, intervals) between scale values.

A paradoxical consequence follows from Carnap's differentiation of two distinct kinds of distance. If the temperature distances are not metrical magnitudes, yet are measurable in the narrow sense of the word, measurability cannot be considered an essential characteristic of metrical magnitudes. We must then presuppose that there could exist metrical magnitudes that are not measurable in the narrowest sense of the word, and non-metrical magnitudes that are measurable in this way.

This paradoxical conception is conditioned first of all by the fact that the concept of a metrical magnitude is identified with the concept of an additive magnitude. If the condition of additivity is therefore a necessary precondition of measurability, every magnitude which does not satisfy it is to be regarded as non-measurable. This is, however, in conflict with the observation that non-additive magnitudes, for example, temperature, can be measured too. Beyond any doubt, Carnap is right when he distinguishes two different kinds of distances, yet he is mistaken in the manner of their interpretation. In its primary meaning, distance, as we have already mentioned, is a geometrical concept, specified in the theory of measurement by the metrical magnitude of length. In its secondary meaning, in the theory of scales, distance (distance function) is a relation between two scale values, either between an arbitrary scale value and the scale zero, or between two immediately adjacent scale values – a relation which is numerically represented as their difference expressed in absolute values.

In order to differentiate between these concepts which differ in meaning, we consider it convenient to talk, in the first case, about *distance*, while in the second, about *scale distance*, or *scale distance function*.

The scale distance should not be identified with the so-called temperature distance, either. The temperature distance, which expresses nothing but the difference between temperatures on one or various temperature gauges, is an empirical concept. The scale distance, which refers to the form of a scale, is a theoretical concept just as the scale zero is. This notion is connected much more with the measurement unit than with the scale zero – a fact that is often consciously forgotten in the theories of extraphysical measurement, perhaps for conceivable reasons. If we compare the forms of two different scales, we must compare their scale values above all with regard to the size of the measurement units of the respective metrical magnitudes. Only then, if the scales do not have the same origin, as happens exactly in the case of measuring temperature, must we also take into account the mutual relation between their scale zeros and find some common point of reference with regard to which this comparison might be homogenized. This homogenization, which is being materialized for scales with different origin precisely with the help of the scale distance between a certain scale value and the scale zero, enables us to find out whether their numerical difference is equal or not, and under what transformational conditions we could convert them into each other. Of course, we also encounter a similar homogenization in the case of comparing scale values for one and the same magnitude, which is measured by means of measurement units of different systems.

Furthermore, what is the relationship between the concept of a scale distance and the concepts of additivity and metric connection? If we realize that this concept has been introduced as a fundamental characteristic of interval scales, which should justify the measurability of non-additive metrical magnitudes and non-metrical magnitudes, then the scale distance will be, so to speak, *ex definitione*, non-additive. In this, it also differs from the concept of distance which is undoubtedly an additive one. Yet, how should we explain the conflict between the non-additivity of the temperature distance and the additivity of the numerical difference between pairs of the corresponding scale values? Apparently, the most simple explanation would consist of regarding scale distance also as additive. However, this is rejected,[18] not only because of considering temperature distances at the same time, but chiefly since such a position would be in conflict with the reasons for which the interval scales had been introduced. Therefore, there is a search for other ways out from this situation.

Either we shall not regard non-additive distances as metrical magnitudes – a position held consistently yet erroneously by Carnap – but then we could hardly explain why these distances are still measurable in the narrowest sense of the word; or we can regard non-additive distances as metrical magnitudes, but then we face the problem of how to justify the assumption of the correlation between non-additive empirical distances and additive numerical differences between the pairs of scale values. If we can derive from the properties of numbers (with possible modifications or without any further restrictions) relevant information concerning the measured magnitudes, as it is particularly assumed by the adherents of the formal theory of measurement, then we must admit that we are indeed able to find empirical counterparts of numerical properties, relations, and operations. Or, alternatively, we must concede that not all properties of real numbers, for example, additivity, will be empirically realizable. However, then we must reject the presupposition of a homomorphism (or isomorphism) between a real-number relational system and some empirical relational system.

These difficulties are not removed by the theory of metric connection, either. The unrealistic nature of this conception is demonstrated by the fact that the axiom of reflexivity [19]

$$a \circ a = a, \tag{4.3–7}$$

required for the operation of constructing the midpoint, is numerically meaningful only for a unique numerical value of the variable a: Either for 0, if we interpret the sign $\circ$ as the symbol of addition; or for 1, if we interpret it as the symbol of multiplication. Yet this axiom is no more meaningful in its numerical interpretation if we understand the sign $\circ$ as the symbol of subtraction, as it is required by the axiom of bisymmetry (4.3–1), unless we regard this axiom as only a trivialization of the axiom of associativity for addition.

For the practical utilization of this conception, the empirical interpretation of individual axioms and an operational realizability of the operation of constructing the midpoint are even much more important. If we take into account the existence of non-additive metrical magnitudes, the difference between non-metrical and metrical magnitudes is conditioned by the existence of a measurement unit and not by the possibility of finding some empirical counterpart of the numerical operation of addition. This necessary precondition of measurability, in the narrowest sense of the word, must also be satisfied in the case of a generalized conception of metric connection. This also holds analogically for a scale distance (distance function), the existence of which is realized by means of a "common and constant unit of measurement."[20] Under this presupposition, all immediately adjacent scale distances will be constant and their mutual relations will be dependent upon the origin of the scale, irrespective of whether this is an *arbitrary* or a *natural* scale zero.

If we take the origin of a scale only as a point of reference, with regard to which the scale values are oriented, and if we identify, for reasons of easier calculation, the scale zero with the numerical zero, then we can deal only with the measurement unit itself. The scale distances will then be uniform in all cases. Between any two arbitrary, immediately adjacent scale distances they will always be equal to the measurement unit or to some of its portions or multiples. Of course, the concept of a scale distance was introduced in the theory of extraphysical measurement precisely for the magnitudes for which no appropriate measurement unit could have been found thus far. The assumption, under which the conception of a scale distance would be meaningful, thus contradicts the reasons for taking on this geometrical concept.

For metrical magnitudes, the corresponding metrical scales will be *uniform*,[21] in the sense that a progression of scale values will be unambiguously delimited by the multiples of the measurement unit. Also, for this reason, a distance function for a determination of the forms of these scales is superfluous. However, this function has a great significance for the definition of the form of non-metrical scales that are *non-uniform*. Yet, since in this case we cannot stipulate any measurement unit, the distance function cannot be realized. There is an obvious conclusion inferred from this: either we have at our disposal a measurement unit and then the scale distance – as well as the metric connection – is unnecessary for the stipulation of the form of scale; or we do not have it at our disposal and then this function, or this connection, cannot be objectively interpreted empirically and exactly realized numerically.

For the uniform metrical scales

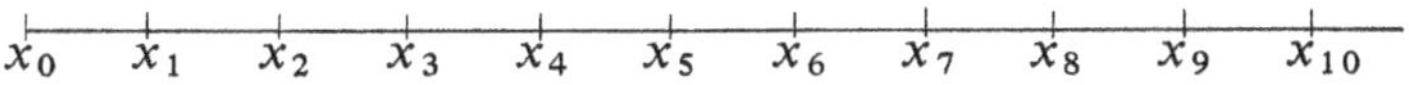

these two conditions obviously hold:

$$d(x_0, x_1) = d(x_1, x_2) = d(x_2, x_3) = \cdots = d(x_m, x_{m+1}) \quad (4.3\text{–}8)$$
$$d(x_i, x_{i+1}) = \mu, \quad (4.3\text{–}9)$$

where μ is a measurement unit, or, respectively, fractional or multiplicational unit. The uniform metrical scales are therefore *equally spaced*.

For the non-uniform metrical scales,

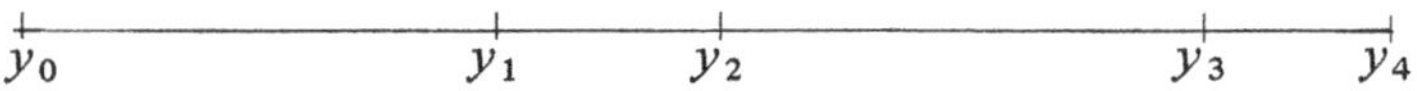

which are not equally spaced, these conditions are not satisfied. Nevertheless, it makes sense to require for these scales some scale distance function

$$d(y_i, y_{i+1}) \gtreqless d(y_j, y_{j+1}),$$

which, however, will not be realized by means of a "common and constant unit of measurement."[22]

In order to be able to apply fundamental numerical operations to scale values, we must assume that the respective scales are uniform. This can, however, only be guaranteed for metrical scales inasmuch as their form is determined by a measurement unit and a scale zero. Because these two conditions are necessary and sufficient, the condition of a scale distance is *redundant*. This fact cannot be bypassed by virtue of replacing the measurement unit, while determining the form of uniform metrical scales, by a distance function satisfying only the condition (4.3–8). At the same time, if the condition (4.3–9) is not also satisfied, then the equal spacing (achieved, for example, with the help of the operation of constructing the midpoint) is not warranted.

It is the measurement unit, and not the nature of the scale origin, which is further decisive for the differentiation of uniform and non-uniform scales. From this standpoint, one should also judge the usual claim[23] that the scale values on interval scales could not be regarded as multiples of some other scale value. The contention that arithmetical operations may be performed only with the intervals of scale values and not, however, with the scale values themselves, is, for example, justified by the claim that – let us say – the temperature 64°F cannot be viewed as the double of the temperature 32°F. On the Celsius scale, the values 17.8°C and 0°C correspond to these values. The operation of multiplication is therefore admissible only when we compare pairs of scale distances on the same scale as referring to its specific origin. We should, however, also arrive at a similar conclusion in applying arithmetical operations to the scale values of any other metrical magnitude, relativized to different points of reference.

This argument is, therefore, only seemingly conclusive and convincing. Two different positions are mixed up in it: the relation of temperature scales to distinct scale origins on one side, and the properties of a certain temperature scale itself on the other, where in one case the scale zero is identified with the numerical zero, while in the other case, it is not.

The uniformity of metrical scales is an essential characteristic of these scales with regard to non-metrical scales, the spacing of which does not satisfy any familiar law-like regularity. Metrical scales need not always be uniform. Their scale values could, for example, be transformed so that we can express each of them by the logarithm of the respective scale value. The transformation of a uniform metrical scale into a logarithmic one, which is not equally spaced, is of course already a secondary operation which refers to scale values and not immediately to the size of the measured metrical magnitudes.

5. QUANTIFICATION

After having exposed the concepts of measurement, magnitude, and scale, there still remains the task of considering the content and extension of the concept of quantification. In the explication of this concept, we meet similar conceptions as in the case of the concept of measurement. If we understand under the concept of quantification a transition from classificational concepts to metrical concepts, or any other procedure by means of which empirical variables are associated with numerical variables (yet not replaced or reduced, as is sometimes unjustifiedly claimed), we may analogically, according to the nature of numerical assignment, discriminate between three different levels of quantification. These extensionally distinct species, operationally characterized by the procedures of numbering (numeral designation), ordering and measurement, correspond then with the concept of measurement in the broadest, wide and narrow sense of the term, respectively. The procedures of numbering and ordering, sometimes only the procedures of ordering and yet, at other times, measurement procedures too, are usually designated in the methodological literature of extraphysical measurement by the term *scaling*.

Quantification in the empirical sciences comprises, not only measurement, but also another very important procedure – the operation of counting.

5.1. SCALING

The term *scaling*, sometimes regarded as a synonym of the term *scale*,[1] is used chiefly with two meanings. First, scaling is conceived as a collection of methods, procedures, or techniques which facilitate the constitution of any scale or only some non-metrical scale.[2] If, besides numerical mapping, the assignment of numerals is also admitted, then not only ordering but also numbering fall into the extension of the concept of scaling. Secondly, in psychology, sociology and other social-scientific domains, by scaling one signifies the actual process of measurement, more precisely, the process of the so-called qualitative measurement that only leads to scale values of non-metrical scales.[3]

For the differentiation of both these meanings it will be convenient to introduce, as in the explication of the concept of measurement, a certain terminological convention: in the first case we shall speak of *scaling procedures* or *scaling techniques*, whereas, in the second case, of *scaling*.

Scaling procedures play an analogical role in extraphysical measurement as in physical measurement. Yet, in contradistinction to operational procedures used in various measurements in physics, these procedures are much more indeterminate and arbitrary. This holds especially for the procedure of numbering, applied either to individual objects or classes of objects, for example, telephone holders, travel passports, houses, streets, apartments, shoe or dress sizes, catalogue items, and the like.

Numbering (numeral designation) has actually the same function as name designation. Therefore, no numerical relation can be presupposed between numerals used only as specific names. If, for example, a goalie of a hockey team is designated by the numeral 1, this does not mean that he became a member of this team sooner than a defender designated by the numeral 9. If one of the forwards is designated by the numeral 6 and another by the numeral 12, it cannot be inferred from this conventional designation that one of them is twice as good as the other. However, one cannot deny that numeral designation has many advantages: it is more transparent, has greater informative value, makes remembering the names of enumerated objects easier, and enables us to extend their amount in an entirely arbitrary manner. If a person is located, for example, in Prague on *Pařížská ulice* (Paris Street), then he need not yet know – if the orientation plan of the city is not at hand – where *Žatecká ulice* is. On the other hand, if the person is on East 26th Street in Manhattan, he will have no difficulties in finding East 28th Street. This difference between name and numeral designation should not, of course, be overrated, for the letters of the alphabet also constitute a certain progression of symbols. Under no circumstances, however, could one regard numbering as a kind of measurement. It namely follows from such a view, advocated by some methodologists of extraphysical measurement, that with the help of this specific naming, we might already arrive at quantitative data.

No wonder, therefore, that the concept of scaling is sometimes interpreted with the same meaning as the concept of measurement and that this interpretation is also being transferred to the explication of the concept of scale. In justifying this conception, the point of departure is provided either by the concept of measurement, if fundamental (basic), derived, and associated scales[4] are distinguished analogically, in accordance with Campbell's classification or its modification,[5] or by the concept of scale, if there is talk, in accordance with Stevens' classification of scale types (Section 7.1), for example, about an interval measurement or about a nominal level of measurement.[6] This conceptual contamination might be explained by an inexact, or perhaps abbreviated, manner of expression. In the former case, one should say explicitly, "scales that we have acquired as a result of a fundamental, a derived, or an associated measurement," while in the latter, "measurement which leads to a constitution of an interval scale." However, the real reason for this confusion of the concept of scaling and measurement (scale and measurement)[7] is again the effort of some methodologists of extraphysical measurement to conceive the concept of measurement, and thus also of quantification, in the broadest possible sense of the term.

As we have already mentioned, we encounter this conception particularly in Stevens' doctrine where the rule of assignment of numerals is regarded, not only as an essential mark of the definition of the concept of measurement (Section 2.2), but also as an essential characteristic of the concept of scale.[8] We also find a similar position, in a modified form, in other authors. Thus, for example, Ellis[9] makes the rule of assignment of numerals more precise by adding the requirement of determinacy, according to which the same numeral ought to be assigned to the same things under the same conditions, as well as the requirement of non-degeneration, according to which different numerals ought to be assigned to different things, or to the same thing but under different conditions. In doing this, Ellis identifies the scale of measurement with the existence of such a rule which is, at the same time, regarded as a specific difference in the classical definition of the concept of measurement.

The problem of whether we can quantify some concept or not indeed does not depend on the assignment of numerals, which may at any rate always be conventionally materialized, but on the objectively existing aspects of the corresponding objects, phenomena, or processes. To comprehend the assignment of numerals as an instance of quantification means to conventionalize this concept to such a degree that there would simply cease to exist any difference between what is measurable and what is not measurable and what is quantifiable and what is not quantifiable.

It must be stated, however, that some methodologists of extraphysical measurement hold a much more realistic position[10] when they stipulate the extension of the concepts of measurement and quantification. These authors talk about scaling only in connection with the construction and use of topological (non-metrical) scales, refuse to include the procedure of numeral designation among scaling procedures and regard the identification of the concepts of scaling and measurement as inadmissible.

In harmony with this viewpoint, which we have already defended in the exposition of the concepts of measurement and magnitude, besides the terms *measurement in the broadest sense of the word, measurement in the wider sense of the word* (measurement on a lower level) and *measurement in the narrower sense of the word* (measurement on a higher level), which have been used so far, we shall also use synonymously the following terminological equivalents: in the first case we shall speak about *numbering*; in the second, *scaling*; and in the third, *measurement*.

Scaling may then be understood as an independent method, which is tied up on the ontologico-gnosiological level with weakly (not, however, with strongly) quantitative aspects of objectively existing objects, phenomena, and processes, on the empirico-mathematical level with non-metrical magnitudes, and on the theoretical level with topological concepts. Alternatively, scaling may be understood as the inevitable first stage of measurement which delimits topological conditions of metrization. Although these alternatives are not mutually incompatible, we shall give preference to the former. Naturally, we must also adjust the interpretation of scaling procedures to this conception. The scope of these procedures will be restricted only to such operational proceedings which facilitate the construction of non-metrical scales. All that we have established for the terms *scaling* and *scaling procedures*, we shall also analogically refer to the derived terms *to scale* and *to re-scale*.

What then is the relationship between the concepts of numbering, scaling, and measurement on one hand, and the concept of quantification on the other hand? If a demarcation line is being drawn only on the operational level, i.e., if a distinction is made between the assignment of numerals and numerical assignment, only the concept of numbering[11] is taken out from the extension of the concept of quantification. However, if we consider as much more essential the differentiation and mutual bonds between qualitative, weakly and strongly quantitative aspects of objectively existing objects, phenomena, and processes on the ontologico-gnosiological level, then, on the basis of this primary criterion, we also have to eliminate the concept of scaling from the extension of the concept of quantification.

Following the intentions of this conception, we shall regard, for the time being, only measurement as an instance of *quantification*. Numbering (numeral designation) might then be taken for *pseudo-quantification* and scaling (numerical ordering), for *quasi-quantification*.

5.2. COUNTING

Numerical mapping, which represents strongly quantitative aspects of real objects, is not restricted only to measurement, but also encompasses counting that must be regarded as another, even more elementary, form of quantification.

Under *counting* we understand an empirico-mathematical method with the help of which one can find out the amount or number of elements of some collection (class, set) of objects having at least one common property. Countable objects are collections of discrete, well-discernible individual objects, phenomena or processes, or collections of such objects which we can treat as if they were discrete.

By means of counting we might, for example, specify the number of persons who had some infectious disease in their youth, since individual persons are easily distinguishable, just as it can be objectively ascertained that they once had this disease. However, we already face difficulties when we try to count drops of water, unless we stipulate by some convention what we understand as a "drop of water", and use some device, say, a dripping-bottle, that will divide a stream of water for us into relatively equally large portions. If we should give the number of men and women in some population, we might easily establish an unambiguous criterion for the differentiation of elements of both these subsets. However, if we are to ascertain in a certain collection the number of qualified workers, artistically valuable novels, mathematically gifted children, and the like, although we can find out the number of individual elements in each group, the result of counting will not be exact if it cannot be decided, on the basis of additional criteria of content, which elements we are to include into the given class and which into the complementary class.

Furthermore, counting presupposes that the counted elements of the given collection objectively occur in a certain temporal progression, if, for example, we count the number of pulses or the number of vehicles proceeding through a cross-road, or that we could register them gradually, if they occur simultaneously (for example, if we determine the number of books in a library, the number of objects in a room, and the like). A collection of objects, the number of which we want to ascertain, need not be qualitatively homogeneous. Among them there may be, not only specifically identical objects, but also very heterogeneous things, for example, objects located in a lost-and-found department. Nevertheless, such collections of objects as, for example,

{apple, apple, apple, apple, apple, apple}
{Europe, Asia, Africa, America, Australia, Antarctica}
{umbrella, pliers, cage, dog, briefcase, man's left shoe}

are similar with respect to their number; as a matter of fact, in the given case they are the same. If we mutually compare them so that we assign exactly one element of another collection to an arbitrary element of one of these collections and vice versa, we need not yet use numbers at all. Of course, in this way one might ascertain only whether the collections considered have an equal number (are equinumerous), or whether one of them has a greater or smaller number (is more or less numerous) than others. The common property of these collections, *to have a greater number than/to have an equal number as* (to be more numerous than/to be equally numerous as), is only a comparative notion under these conditions. From this comparative concept we come to the metrical concept of *number, numerousness*, if one of the compared collections of objects is a set of non-negative whole numbers.

Since the ascertainment of the number of elements of some collection of objects with respect to the number of elements of another collection is also possible without the use of numbers, counting is often held to be an empirical operation that, from the operational standpoint, depends upon numbering.[12] Yet, counting is also regarded as an empirical operation when numerical mapping has already been used, for the character of an empirical operation is also assigned to such a superior procedure of measurement. From the assumption that counting is a case of measurement (either its most basic or only its specific kind, namely the measurement of numbers), it is inferred that a number – more precisely, a "physical" or "empirical" number, amount, or numerousness – is a physical property of empirical classes of objects, and thus an unnamed physical magnitude that is fundamentally measurable.[13]

One cannot agree with this conception of counting. Consider first of all the difference between numbering and counting. Numbering might also be applied to objects of one collection, whereas counting always refers at least to two collections: either to arbitrary collections of objects, if we deal only with the comparative notion *to have a greater number than/to have an equal number as*, or to an arbitrary collection of objects and a set of non-negative whole numbers, if we refer this operation to the metrical concept of number, numerousness. In numbering, we characterize a given collection qualitatively: we state *which* elements are contained in it. By counting, we delimit a collection quantitatively: we ascertain *how many* they are.

Numbering is not tied up with the use of numbers, but only with naming of individual elements of the given collection of objects. By virtue of this, this obviously *empirical* operation differs, on one hand, from the *empirico-mathematical* operation of counting, which is based on a one-to-one relation between some collection of objects and the set of non-negative whole numbers and, on the other hand, from calculation (computations), which comprises various *mathematical* operations with numbers. Counting should also be considered an empirico-mathematical operation when we admit, besides the mathematical concept of number, a "physical" notion of number too, for in this case it must also be acknowledged that one deals here with a mapping of a certain collection of empirical objects into the set of numbers.

Secondly, to what extent may one meaningfully speak about "empirical numbers" and regard them as fundamentally measurable magnitudes? Since we have already commented upon the mutual relationship between numbers and magnitudes (Section 3.1) and critically evaluated Campbell's nominalistic comprehension of mathematics (Section 2.2), it is not necessary to justify the illegitimacy of this conception in greater detail. It is evident why we encounter it exactly at the explication of the concept of counting. If counting is an instance of measurement and if measurement refers to magnitudes, then that which is the object of counting must also be considered a magnitude. There exists, however, an even weightier reason. If one could regard physical numbers, or a number of elements of a certain collection of objects, as a specific magnitude that is even fundamentally measurable, then the possibility of measurement is justified almost without any restriction. The specification of the number of elements of an arbitrary collection of homogeneous, and also very heterogeneous, objects by means of counting, inasmuch as these elements might be regarded as individual objects, is namely realizable under all circumstances.

Regardless of the assertion that counting is a specific case of measurement, it should not be overlooked – on the basis of the above-mentioned explication of the concept of counting – that there exist remarkable distinctions between these procedures. Counting implies only mapping onto the set of non-negative whole numbers, while measurement is tied up either with mapping onto the set of rational numbers (if we consider metrical magnitudes, the values of which are results of an experimental measurement) or onto the set of real numbers (if we also take into account computations in working out the experimentally accessible numerical values). Measurement presupposes ordering, whereas in counting it does not matter in what order we count the individual elements of a given collection. By virtue of measurement, we quantify homogeneous properties of objects, phenomena, and processes, while by means of counting we ascertain only the amount of individual elements of collections of objects that are often qualitatively very heterogeneous.

An advocate of the criticized conception need not, however, acknowledge these distinctions as relevant. He might point out that certain metrical magnitudes are also discrete, so that they might be regarded as classes of individual elements. He might emphasize, too, that the set of non-negative whole numbers is contained in the set of real numbers – from this he may even infer a confirmation of his own position. Yet, even if he speaks about the measurement of the number of elements of a class, about the measurement of the numerousness of these elements by means of counting, or about a measurement of "physical" numbers, this advocate cannot explain away the non-existence of the essential component of measurement – the measurement unit and, consequently, the non-existence of a metrical scale as well.

B. Ellis [14] is well aware of this fact, although he also defends the position according to which a "physical" number is a magnitude, yet at the same time, he doubts that it is at all suitable to regard counting as measurement, and that it is meaningful to talk about a measurement scale for this magnitude, if we cannot stipulate any measurement unit for it. Likewise, Suppes and Zinnes acknowledge that the number of elements in a given collection is unambiguously defined by counting, yet without having at our disposal an arbitrary measurement unit or a scale zero. However, they conclude from this that counting is a specific sort of measurement which is carried on with the help of the so-called absolute scale.[15] Because they distinguish this scale from ratio and interval scales, apparently they acknowledge that a natural measurement unit and a natural scale zero can be established for it. Stevens [16] includes among the ratio scales the scale of numbers which he, at the same time, interprets as a scale of the numerosity of empirical collections of objects. He even declares this numerical scale – the scale of cardinal numbers used in counting – to be the most fundamental sort of ratio scales. Contrary to Suppes and Zinnes, he must therefore assume that some arbitrary measurement unit could be found for it.

These mutually incompatible views demonstrate very convincingly that if one subsumes the concept of counting under the concept of measurement, regards also number, numerousness, as a metrical magnitude, and also includes the numerical series of non-negative whole numbers (i.e., the *numerical scale* itself) among metrical scales, then one has to face many difficulties which follow from such views.

What, however, is the measurement unit and the scale zero of this absolute scale? Its measurement unit can only be number 1, referring to arbitrary individual elements of every non-empty set of objects. Yet its scale zero cannot be the number 0, for this number, too, indicates the numerousness of elements of the counted set. If we restrict counting to only non-empty sets, the scale zero of the absolute scale would again have to be number 1. However, then there would be no difference between the natural origin and the natural measurement unit of this specific scale. Hence, the absolute scale does not satisfy one of the conditions which are imposed on the form of metrical and non-metrical scales.

In fact, the unit of the absolute scale should not be understood as a measurement unit, but only as the *unit of counting*. The unit of counting, though referring to various individual elements of some set, is always an unnamed number. In contradistinction to measurement, in counting we abstract from the qualitative specification of the counted objects. As to their number, all equally numerous sets are identical, even if they consist of qualitatively entirely heterogeneous objects (take, for example, sets of six villages, six continents, six theories, and the like). To what extent is the unit of counting *natural*, as assumed by Suppes and Zinnes, or *arbitrary*, as Stevens thinks? Its "naturalness" obviously rests on the fact that it always refers to a *unique* individual element of the given set, and its "arbitrariness" is given by another fact that this individual element may be *any* individual object. In this, therefore, the unit of counting is again essentially different from the unit of measurement.

If counting is held to be a sort of measurement and, consequently, the unit of counting is regarded as a measurement unit, then by means of counting, we might also measure non-metrical magnitudes for which we do not have at our disposal any measurement unit. Since we can, for example, count the number of persons visiting a theatre, the number of persons having a positive attitude to sports, and the consumption of paper per capita, etc., many methodologists of extraphysical measurement derive from this a conclusion that, by virtue of counting, one could measure theatre attendance, attitude toward the sport, cultural level, and the like. Should this indeed be the case, we might simply measure anything and, in principle, every general concept, the extension of which always covers more than one object, would have to be rendered as a metrical, quantitative concept.

By means of counting, one can actually find out only the number of elements of some collection of objects. By virtue of this initial form of quantification, we acquire basic quantitative data, yet from these alone we cannot derive any further conclusions that would be immediately related to the qualitative aspects of the counted objects, phenomena, or processes. This also holds for such concepts as, for example, accident rate, casualty rate, birth rate, or death rate. If we understand these and similar notions only as being synonymous with the concepts "the number of accidents", the number of casualties", "the number of born children", or "the number of dead persons", i.e., as quantitatively defined concepts (not, however, as metrical concepts in the sense of our introductory exposition), then on the basis of the ascertained quantitative data, we can say that during a certain period there occurred within a certain population so and so many accidents or casualties, that so and so many children were born, and so and so many persons died. Inasmuch as we wish to express something more by using these concepts, the quantitative data acquired by counting constitute only one of their symptoms. These quantitative data, providing material for further statistical inquiries, display the character of the so-called indicators (indices) which, however, do not represent metrical magnitudes in the sense of the theory of measurement.

Similarly, one cannot regard as an instance of the empirico-mathematical method of measurement, the so-called measurement of correlation which is essentially only a statistical elaboration of quantitative data or, more precisely, an *enumeration* of various statistical characteristics. Here we have another reason why the methodologists of extraphysical measurement subsume counting under measurement, overlooking the essential difference between the primary use of the empirico-mathematical method of measurement and the secondary statistical evaluation of quantitative data, irrespective of whether these were obtained by measurement or by counting. Even less could one regard a statistical "measurement" of correlation between different indicators, acquired by counting, as a conclusive argument for the justification of the applicability of measurement procedures within the frame of a behavioral or social measurement.

Besides the view that counting is a specific sort of measurement, we also encounter a contrary position, according to which measurement is reduced to counting, namely to counting the number of measurement units contained in the measured values of metrical magnitudes.[17] Although this contention, which finds its support instead in the practice of physical measurement, is less controversial than the opposite position, it nevertheless does overlook the difference between the primary and secondary use of the operation of counting and, consequently, leads to an oversimplified comprehension of the problems of measurement.

It has already been emphasized that counting determines the number of individual objects of a collection and, for this reason, refers primarily to discrete entities and individual objects and not, however – unlike measurement – to their continuous properties. If, with the help of the measurement unit, we decompose the measured magnitude into a series of individual elements of the same size, we might also secondarily apply counting to this specific sort of discrete objects and regard this operation as an auxiliary component of measurement procedures. However, one must not infer from this, as a general conclusion, that measurement is reducible to counting.

At the end, it will be appropriate to answer concisely one further possible objection. If we understand counting as a sort of quantification, do there not follow from this the same consequences as from the criticized position, according to which counting is a specific sort of measurement? If we remind ourselves of our differentiation between pseudo-quantification, quasi-quantification, and quantification, then it must be quite obvious that in our interpretation one cannot – in contradistinction to counting and measurement – regard numbering and scaling as kinds of quantification. And it is exactly this respect which constitutes the principal difference between both conceptions.

6. THEORY OF MEASUREMENT

One can speak of a theory of measurement in various connections. This situation is caused not only by our peculiar ways of interpreting the concept of measurement (we might interpret it in a wider or a narrower sense of the term), but also by our intentions concerning the extent of the theory (it matters to what extent we wish to go in drafting the theory). After all, we might consider a general theory of measurement, specific theories of measurement (in particular, those of physical and extraphysical measurements), and theories of a certain kind of measurement, for example, fundamental measurement, which may eventually be specified with respect to a definite scientific discipline. Furthermore, we might consider a theory of measurement of some metrical magnitude (i.e., mass), a theory of measurement procedures on a general level, or only a theory of measurement procedures that are either applied in measuring one magnitude or used with the aim of attaining a certain interval of scale values. When we constitute a theory of measurement, whatever its scope and aim, we may emphasize conceptual, methodological, and operational aspects of measurement, its empirical and mathematical characteristics, or we may consciously concentrate only on an analysis of some of these. These possible approaches are, of course, always influenced by general methodological conceptions and philosophical views and, thus, must lead to distinct, sometimes even quite contradictory, results, in spite of all the agreement on the choice of the themes.

The classical theory of measurement, anticipated by von Helmholtz and elaborated by Campbell, contains above all the formulation of the so-called laws of measurement, namely, the law of ordering, the laws of addition, and the law of a standard progression, as well as an analysis of two basic sorts of measurement – fundamental and derived. In the theory of measurement elaborated by Carnap, Hempel, Stegmüller and other authors, depending upon the differentiation of concepts into classificatory, comparative, and metrical, the basic components of the classical theory of measurement have been enriched by the axiomatization of the necessary and sufficient conceptual preconditions of measurability in terms of establishing the conditions of metrization. In Stevens' position, the theory of measurement is essentially interpreted as a theory of scales of measurement which is based primarily on the criterion of the transformational properties of the scale types and admissible statistical operations with empirical data, as contingent upon the type of scales. Under the influence of von Neumann-Morgenstern's axiomatic theory of expected utility and Tarski's doctrine of the concept of model, in the formal theory of measurement which was worked out by Suppes, Zinnes, and other methodologists of extraphysical measurement, the main attention is focused on the analysis of formal properties of different systems of measurement or measuring structures and on the derivation of two basic theorems – the so-called representation theorem and the uniqueness theorem.

In this chapter, the goal of our exposition will be a critical analysis of only a few basic components of the general theory of measurement, particularly classifications of kinds of measurement, conditions of metrization, and problems of representation, including the representation theorem. We shall treat further questions only in our analysis of the theory of scales. In both cases it will be shown that there exists considerable agreement between different approaches to the theoretical foundations of measurement and that many problems are often solved only in an alternative manner.

6.1. REPRESENTATION THEORIES OF MEASUREMENT

All general theories of measurement advocated by the methodologists of extraphysical measurement – irrespective of whether these theories follow the classical doctrine of Campbell or are the so-called liberalized theories [1] – explicitly or implicitly rest on the presupposition of a homomorphism (or isomorphism) between some empirical relational system and a certain numerical relational system. Because, by virtue of this presupposition, a certain representative function is quite generally assigned to the numerical mapping, these theories are designated as the so-called *representation theories* of measurement.

Then, the difference between particular variations of the representation theory of measurement lies in the way in which they interpret the nature of the numerical mapping, how they understand the mutual correspondence between numerical and empirical relations and operations, how they justify the possibility of this mutual correspondence, and what consequences they draw as to the representation of empirical characteristics of the measured magnitudes, particularly with respect to the nature of the resulting scale values.

Depending on the extent to which the concept of measurement is subjected, the representational function is assigned to numerals, ordinal numbers, or only to cardinal numbers. Analogically, the representational nature of numerical relations and operations is also interpreted in these distinct ways. Some writers hold the view that numerical relations and operations must represent the corresponding empirical relations and operations only in the case of a fundamental measurement, while others also require this from the other kinds of measurement, at least to a certain degree. According to the application of measurement to strictly extensive, extensive, or quasi-extensive magnitudes, the requirement of correlation between the structure of the empirical and numerical relational systems is restricted either to relations and operations, or only to relations.

Of course, it is not enough if the representation theory of measurement merely claims that numbers are assigned to empirical objects in such and such a way and that definite relations between numbers and operations with numbers correspond to certain relations between them and operations with them. This theory must also theoretically justify the conditions under which this assignment is not only possible, but also empirically meaningful and operationally realizable. When one justifies the factual existence and meaningfulness of this correlation, which is defined by the rules of correspondence between empirical and numerical relational systems and between empirical facts and numerical expressions, one might, in principle, start from some numerical relational system and look for a suitable empirical model for it, or, to the contrary, formulate – on a more general level, or on a specific one – formal characteristics that must be fulfilled by the investigated empirical relational system if we are to find for it the corresponding numerical relational system.

This basic problem of every general theory of measurement is resolved in two ways: by the stipulation of the *conditions of metrization* for objects of the empirical relational system, or by a positive solution of the so-called *representation problem*, i.e., by a derivation of the respective representation theorem. In either case of this theoretical founding of the measurability of a certain property or magnitude (with essentially the same intention), one usually starts from an axiomatic formulation of both relational systems, though, as we assume that it is quite sufficient to establish the conditions of measurability in an axiomatic form only in one case and then supplement it with the respective rules of correspondence. The derivation of the representation theorem, just like the metrization, are specified at the same time with respect to different sorts of measurement or different species of magnitudes.

Because the resulting numerical values, assigned in the process of measurement to empirical elements, are unique only up to a certain measure, for example, in the case of measuring strictly extensive magnitudes up to the choice of measurement unit, another basic problem of the representation theory of measurement is the so-called *problem of uniqueness*. Within this problem one deals with the ascertainment of the measure of this uniqueness and with establishing the conditions under which different scale values could represent the same weakly or strongly quantitative aspects of the measured properties or magnitudes.

The problem of uniqueness might again be resolved in two ways: either for every sort of numerical mapping of an empirical relational system there is derived a specific theorem of uniqueness, defining the transformation conditions under which scale values remain invariant, or, with the linkage to a classification of scales, there are given invariant transformations under which the scale form does not change. Since the invariance as to the admissible transformations is a primary basic criterion for the differentiation of scale types, it will be more expedient to inquire into this second basic problem of the representation theory of measurement only in connection with the theory of scales.

6.2. KINDS OF MEASUREMENT

Before we approach the problems of topologization and metrization, as well as the discussion of the problem of representation, it will be necessary to analyze various classifications of the kinds of measurement. These classifications, usually dichotomic, not only refer to the systems of measurement, but also to measurement procedures. They are also combined with the classification of magnitudes and metric scales.

In all theories of measurement we encounter the differentiation between the so-called fundamental and derived measurements, which had already been anticipated in Helmholtz's analysis of measurement and counting. However, this basic classification is not interpreted in the same way in other theories of measurement.

In the classical theory of measurement, elaborated by Campbell, every measurement which does not encompass any previous measurements is regarded as *fundamental*. On the other hand, measurements that depend on previous measurements are considered *derived*.[2] Analogically, a distinction is drawn between fundamental and derived measuring procedures.

Both these classifications, which are only analogical, start from the familiar observation that some magnitudes can be measured or are measured regardless of other measurements, while other magnitudes are measurable or are measured only by means of already performed measurements, that some measurement procedures are direct, whereas others, in turn, are indirect. From the theoretical standpoint, this conception of fundamental and derived measurement is motivated by the system of physical magnitudes and by dimensional analysis. With regard to practically materialized measurements, it is influenced by the possibility of measuring some magnitudes directly – either by comparison with some series of standard objects, or with the help of suitably constructed measuring instruments – or indirectly, on the basis of a numerical elaboration of the foregoing results of measuring the same or other magnitudes.

As we have already explained in our exposition of magnitudes (Section 3.5), it is furthermore essential for Campbell's theory that fundamental measurement is, in principle, applicable only to strictly extensive magnitudes. Derived measurement, materialized on the basis of numerical or empirical laws, also refers to extensive or quasi-extensive magnitudes.

The significance assigned to fundamental measurement is based on the fact that this sort of measurement provides the point of departure for all other kinds of measurement, that it fulfills all conditions of topologization and metrization, and that it is also operationally realizable.

However, by what means could one justify the priority of fundamental measurement? How can it be explained that a measurement of some magnitudes might be carried out independently of the measurement of other magnitudes? If we are to discuss these problems and to defend the legitimacy of this basic classification of the kinds of measurement, we have first to clarify in what sense the notion of independence is to be understood.

The independence of a fundamental measurement, for example, the measurement of length, which is generally acknowledged as being a basic magnitude in the system of physical magnitudes, evidently may be interpreted only so that, for historical and systematic reasons conditioned by the objective characteristics of this magnitude, the length measurement is a necessary precondition for the measurements of many other magnitudes. Of course, this does not mean that we could measure this metric magnitude independently of the theoretical presuppositions that are linked with the conception of measurement, on one hand, and with the theory of physics, on the other hand, or by disregarding specific operational conditions.

For length measurement there is usually acknowledged a whole range of such presuppositions, which, however, are only partially realizable in reality. Let us mention at least some of them:

(i) the measured object, just like the measuring instrument, in the simplest case – a measuring rod – is an absolutely rigid body;
(ii) the length of the measured object, as well as the length of the measuring instrument, does not change in the process of measurement;
(iii) the results of measurement, despite possible deviations produced by various external influences, fall into a definite numerical interval of values, which represent, with a required measure of exactness, the extensiveness of the measured object;
(iv) the length of some object is empirically accessible, at least within certain limits of its values, and measurable by means of a certain direct measurement procedure;
(v) both objects to be compared – the measured object and the measuring instrument – may be precisely correlated with respect to a given property.

These and other presuppositions refer above all to the operational realizability of the formal properties which are assigned to the empirical counterparts of numerical relations and operations that are necessary for obtaining the scale values of length measurement. Perhaps it is not necessary to stress that such presuppositions are only approximately objectively satisfiable. It is an obvious idealization to assume that the measured object, just like the measuring instrument, are absolutely rigid bodies. The other presuppositions are such idealizations too. Indeed, no real body exists isolated with regard to its environment; no real body is unchangeable and thus it must be changing all the time in the process of measurement. Objectively, a real body is always dependent upon many other conditions, for example, on the change of temperature, pressure, humidity, and the like. Since the measured object and the measuring instrument are different bodies, the congruence of their end parts will always be more or less inexact, depending on the discrimination abilities of the experimenter and, in more complicated cases, also on the sensitivity of measuring instruments and other circumstances connected with a concrete manner of measurement. Equally significant also is the fact that empirical relations and operations correspond to their idealized numerical counterparts only to a certain extent and only within definite limits of the measured values. If we measure length in the extreme values, which are empirically accessible only indirectly, for instance, in the case of measuring interatomic or interplanetary distances, we must take into account the various physical laws referring to other magnitudes as well.

All these facts, especially mutual relations between different properties of the measured objects and the magnitudes that correspond to them, necessarily influence the theoretical and empirical characteristics of fundamental measurement. The independence of a fundamentally measured magnitude with respect to the dependence of non-fundamentally measured magnitudes exhibits itself in the fact that all properties of the measured objects (save their extensiveness, if we consider, for example, the length measurement) may, for theoretical reasons, be rendered constant. Yet, if these magnitudes should remain constant, we must again take into account a whole range of physical laws in which the mutual dependencies between them are defined. The constancy of some conditions in which the measured object occurs must, however, also be analogically preserved when we measure some magnitude derivatively. This dependence on theoretical presuppositions is conditioned in either case by the fact that measurement is not a mere empirical procedure which may be practically applied regardless of its conceptual aspects and independently of the theory of the respective scientific domain.

Can we, therefore, ever claim that the length measurement is objectively independent of the measurements of other magnitudes? Are we not entering a vicious circle that makes the difference between fundamental and derived measurement entirely relative, if we admit, for example, that the length measurement is influenced by the temperature of the measured objects, and if we acknowledge, to the contrary, that the temperature measurement requires previous measurements of length, while in both cases we take other properties of the measured objects, for example, pressure, as constant?

When assessing these questions, we must take into account in the first place objective preconditions of measurability for various kinds of metrical magnitudes from the ontologico-gnosiological standpoint. The priority of fundamental measurement is not an expression of a certain convention, conditioned by the requirement of an operational or theoretical simplicity (as every positivistically oriented methodologist would certainly admit), but a necessary consequence of the historical process of the cognition of law-like regularities in the objective reality, a process which proceeds from the simpler form of the motion of matter to more complex ones. (This, of course, would be rejected by a positivistically oriented methodologist.) From the standpoint of the objective dialectics of the evolution of nature and its reflection in the system of scientific knowledge, it is therefore evident that a classificational system of the kinds of measurement, just like a theoretical system of magnitudes (between which there does not necessarily exist a one-to-one relationship, as we have already emphasized earlier (Section 3.5)) must originate from certain primary measurements and basic sorts of magnitudes.

This justification of the conception of fundamental measurement may be supported by a further argumentation. If we measure some magnitude fundamentally, we also arrive at relatively appropriate numerical results when we neglect taking into account some idealizations referring to real properties of the measured objects or to the circumstances under which they are measured. If we measure, for example, the length of elastic bodies or the length of a certain object in different conditions (say, one time under low temperatures and, another time under high temperatures), the limits for the measured values will only be broader and more variable than ordinarily. Yet this interval of scale values may always be made more precise if we restrict the scope of concomitant conditions or if we introduce suitable correction factors for them; for example, when we measure the length of bodies only in a definite *middle* range of temperature differences or when, in accordance with the material composition of the measured bodies, we modify the empirically ascertained values by the respective thermal coefficient of length expansion.

Nevertheless, no such possibility exists for the magnitudes that are not fundamentally measurable. The measurements of length, mass, or time are not secondary factors of the measurements of density, temperature, pressure, velocity, and the like, but are inevitable initial prerequisites of their measurability, particularly from the conceptual viewpoint. An area or a volume cannot be measured without a prior measurement of length. Density of homogeneous matter, the chief unit of which is kilogram per cubic meter, is defined by a definite relation between volume and mass and thus presupposes measurements of length and mass. The velocity of a uniform motion, determined by the distance s which a body travels in time t, depends on measurements of time and length. Naturally, these dependencies also hold for a relationship between further basic and derived magnitudes. The derived measurement of an electrical charge is conditioned by a fundamental measurement of time and electrical current. Because the dimension of electrical resistance is expressed – in harmony with the SI system – by the dimensional formula

$$[R] = L^2MT^{-3}I^{-2}, \qquad (6.2\text{–}1)$$

the delimitation of this derived magnitude rests on these basic magnitudes: length, mass, time, and electrical current.

The dependence of derivatively measured magnitudes on the fundamentally measurable ones is also preserved in the case when we measure them by means of some measuring instrument, instead of obtaining their scale values by calculation from the numerical values of fundamentally measured magnitudes. Distinct ways of measuring some derived magnitude, materialized with the help of various measuring instruments and diverse measurement techniques, cannot change the character of both these kinds of measurement. Neither could these distinct ways lead to a situation that we might be able to take a derived magnitude for a fundamentally measurable one – as a consequence of the development of instrument technology which enables us to measure the given magnitude directly or in values which have not yet been measurable.

Only an adherent of operationalism might arrive at such a conclusion, since for him there exists no difference between measurement and measuring procedures.[3] Operationalism understands the process of measurement exclusively in terms of an empirical procedure based on observationally controllable methods, connected with the use of measuring instruments, which facilitate the registration of numerical values by reading the data on the registration equipments of material scales. For philosophical, theoretical, and methodological reasons, such a doctrine is, in principle, incorrect. Every empirical procedure – and even more, measurement, of which we have shown that it is an empirico-mathematical method – must always be objectively and theoretically founded. Perhaps it is not even necessary to emphasize that the construction of measuring instruments is quite non-operationally based on objectively existing relations between fundamentally measured magnitudes, represented in the form of theoretically justified numerical laws. The user of some measuring instrument, for example, a driver watching the pointer of the odometer need not, of course, realize this for the entire theoretical work has been already carried out for him by the constructor of the measuring instrument. From a purely practical standpoint of the use of a measuring instrument, one is not at all permitted to draw any relevant conclusions concerning such issues as, what is the principle, under what theoretical presuppositions, and on what real relations between fundamentally measurable magnitudes is the functioning of the instrument based.

From a purely operational standpoint, however, every derivatively measurable magnitude becomes fundamentally measurable, if one succeeds in constructing some measuring instrument by means of which it can be measured directly. Thus, for example, the density of a liquid, inasmuch as its scale values will be obtained by calculation from the measured values of mass and volume, will be conceived by an operationalist as a derivatively measurable magnitude. Yet, so far as it is measured by a pycnometer, he will consider it a fundamentally measurable magnitude. Evidently, it follows from this conception that every change in instrument technology, just like the introduction of distinct measurement procedures applied to different values of the measured magnitudes (which cannot be measured in the same way in the full range of possible values, owing to the objective reasons of their occurrence) must bring about a change in the borderline between fundamental and derived measurement. Since operationalism holds that the concrete way of measuring some magnitude is both a necessary and sufficient condition of its existence, there follows from this, furthermore, a quite arbitrary relativization of the mutual relationship between basic and derived magnitudes in the system of metric magnitudes. This, however, undermines not only the theoretical foundation of physics, but also the stability of the accepted system of measurement units and dimensional analysis. The criticism of these consequences of the operationalist interpretation of measurement does not mean, however – as it has been already stated – that we would comprehend both these, conceptually close, classifications in an absolute sense, independently of the historical evolution of scientific knowledge.

Yet, contrary to appearances, operationalism does not adequately grasp fundamental measurement itself either. Every metrical magnitude, even if it is measured directly by empirically observable procedures, assumes – in accordance with the required homomorphism or isomorphism between both relational systems $\mathscr{E}$ and $\mathscr{N}$ – the possibility of an assignment of numerical values from the realm of real numbers. Nevertheless, by actual, purely empirically performed measurement procedures we may obtain only numerical values from the realm of rational numbers. However, by means of an actual, operationalistically understood, measurement we can never reach real numbers, which are of necessity used in the formulation of numerical laws and without which any theoretical discipline is inconceivable (insofar as it is based on quantitative data obtained by fundamental measurement of metrical magnitudes).

This conflict between empirically measurable values of magnitudes and their numerical values, required for the reasons of calculation, is still exhibited in other ways. Although the set of actually measurable values, i.e., the set of rational-number values, is theoretically denumerably infinite, practically, it is always finite. Each repeated measurement is limited by the finitude of real possibilities under which we carry it on, above all, for reasons of time. Yet the set of all real-number values of a magnitude is nondenumerably infinite; namely, it has the cardinality of the continuum. If the fundamentally measured magnitudes were measurable purely operationalistically, it would be possible to establish, for example, all admissible values for the length measurement, hence its real-number values too, irrespective of the necessary theoretical extrapolations and interpolations of such measured numerical values. Of course, this is not possible in reality. We are compelled to be convinced about this already by such elementary acts of measurement as the measurement of the length of a diagonal of a square, say, with the side $a = 1$ dm. The impossibility of also measuring basic magnitudes by purely operational procedures demonstrates itself even more expressively in the situation where the given magnitude is measured in its boundary values, which are not immediately observable and where we must always deal with calculations on the ground of definite, theoretically justified numerical laws.

While operationalism puts one-sided emphasis only on measuring procedures, the formal theory of measurement, just as one-sidedly, stresses only the function of numerical representation and, naturally, it also assesses from this standpoint the relationship between fundamental and derived measurement.[4] Then every measurement in which the numerical mapping immediately represents some empirical relational system is viewed as fundamental measurement. On the other hand, a derived measurement depends directly, not on some empirical relational system, but on other numerical mappings. The relationship between these kinds of measurement is therefore relativized as to distinct formal structures, for which one might derive different representation and uniqueness theorems.

What is decisive for the formal theory of measurement is the variability of the numerical representation, and not the mutual relations between magnitudes, some of which may be measured only when we have already managed to measure other magnitudes. Therefore, from the standpoint of this theory, the view that there exists only one fundamental system of properties, which adequately leads to a numerical measurement, is rejected and it is claimed that one might formulate different measurement systems, all of which will be fundamental in the sense that they are independent of other measurements. This contention is aimed against Campbell's requirement that fundamental measurement must always fulfill the condition of additivity. Although the representatives of the formal theory of measurement concede that the existence of an empirical operation of concatenation, so far as it occurs, is very valuable, nevertheless they do not consider it a necessary condition of a fundamental measurement.[5] As we have already mentioned (Section 3.5), one might agree with this criticism of Campbell's position. Nevertheless, one cannot say the same either about the point of departure of this criticism or about the manner in which it has been justified.

It can be very easily explained why the representatives of the formal theory of measurement cannot admit, in any case, the applicability of fundamental measurement exclusively to strictly extensive magnitudes. If this measurement is identified with so-called extensive measurement, it follows from this that fundamental measurement is hardly possible in the social sciences. There has not yet been found any appropriate empirical counterpart of the numerical operation of addition for the various properties which are regarded as the object of measurement in psychology, sociology, economics, and the like (for example, loudness, intelligence, different attitudes, utility, etc.). Campbell's doctrine of fundamental measurement is thus in sharp conflict with the reasons which had led to the constitution of this theory of measurement. This theory has been developed precisely for the sake of vindicating the legitimacy of extraphysical measurement and helping to justify the scope of measurability to the broadest possible extent.

Yet fundamental measurement is not the only sort of measurement, hence the requirement of an immediate numerical mapping of some empirical relational system is not a necessary condition of measurability either. Considering the older forms of psychophysical measurement, we might carry on measurement in a similar way in other social scientific domains. This would mean, however, that measurement in these domains would always depend on physical measurements. Evidently, in such a case we could not view extraphysical measurement as an independent sort of measurement. This consequence is again in conflict with the efforts of the methodologists of extraphysical measurement to develop their own theory and methods.

The independence of extraphysical measurement is also emphasized for the reason that it is positively much easier to stipulate formal properties of empirical structures which would correspond, under considerably idealized conditions, to the properties of numerical relational systems, than to investigate content-dependent relationships between physical magnitudes and the corresponding extraphysical magnitudes or purely qualitative properties. This also enables us to regard as fundamental measurement any numerical mapping that leads, say, only to the so-called ordinal scales (Section 7.1). From the standpoint of fundamental measurement, as it is understood by the representatives of the formal theory of measurement,[6] no essential difference exists between this or any other kind of numerical mapping. Each of them, so far as it can be expressed in the form of a homomorphism between empirical and numerical relational systems, is just as good as "methodological example of fundamental measurement as the other."[7]

The discussion of whether it is possible to have fundamental measurement without this being a measurement of strictly extensive magnitudes, steers the attention away from a much more major problem. Under what conditions can we speak in the social sciences about measurement in the strictest sense of the term? The adherents of the formal theory of measurement solve this problem simply by constructing various axiomatic systems for fundamental measurement which do not contain the operation of concatenation as a primitive notion.

Yet the possibility of fundamental measurement cannot be justified only by using formal means. Unless the basic empirical, conceptual, and theoretical presuppositions are satisfied, under which one can meaningfully infer from the properties of numerical relational systems the corresponding properties of empirical relational systems and vice versa, one should not talk about a measurement in the strict sense of the term. Only when this can be shown, may we further consider the question of whether we are or are not dealing with fundamental measurement. If we cannot state that the measurement of some magnitudes presupposes the measurement of other magnitudes, that for some magnitudes one may obtain their scale values independently of the scale values of other magnitudes, while for other magnitudes one might reach them by calculation, it makes absolutely no sense to distinguish fundamental and other kinds of measurement. Yet this can be decided only on the basis of a definite system of magnitudes, explicitly formulated numerical laws and a theoretical basis of the scientific domain involved. Since only the first steps have been taken toward the realization of these necessary preconditions of theoretically founded systems of measurement in the social sciences, every extraphysical measurement, inasmuch as it might be regarded as a measurement and, insofar as we wish to confine it only to the given scientific domain, has essentially the character of a fundamental measurement.

Within the formal theory of measurement,[8] this conclusion also follows from the formulation of the representation theorem, according to which a certain homomorphism may be construed into some numerical relational structure, if the given relational structure satisfies certain axioms. If measurement is then understood as a construction of such homomorphisms, or perhaps scales of measurement, it is evident that what is decisive here is only the numerical mapping, referring immediately to empirical structures, i.e., the construction of a fundamental numerical assignment for the given empirical system. In the initial stages of the evolution of the formal theory of measurement, thought was also given to a derived numerical assignment,[9] yet in the later development of this conception, derived measurement has been replaced by the theories of the so-called *conjoint* measurement.[10] These theories should facilitate the construction of measurement scales for composite objects which preserve their observed order with respect to the given property (e.g., momentum) and where the scale value of each object is a function of the scale values of its components, in the given case – mass and velocity. They should also provide possible methods for the fundamental measurement of magnitudes which do not satisfy the condition of additivity, hence for all sorts of non-extensive measurements, including various kinds of *interval* (difference) measurement, *linear* measurement, the measurement of *utility*, *ordinal* measurement, etc.

Thus, by means of a conjoint measurement, which is considered to lead either to *additive, multiplicative,* or *exponential* representations, the difference should be bridged, not only between fundamental and derived measurement, but also between extensive and non-extensive measurement, as well as between basic and derived magnitudes. Yet still much more is being assumed about this kind of measurement: simply all properties are held to be fundamentally measurable if they satisfy the conditions of ordering, hence non-metrical magnitudes, for instance, hardness, too. However, this does not mean that scaling is also viewed as measurement. Therefore the aim of the criticism of fundamental measurement in Campbell's sense given by the representatives of the formal theory of measurement, is quite obvious. Nothing else is their concern but the possibility of classifying any numerical mapping "according to rules" as a measurement.

A similar conception, explicitly related to metrical magnitudes, may be encountered in the work of Kanger,[11] who distinguishes three basic structures and scales corresponding to them. The first of them represents extensive measurement, while the other two, illustrated by the measurements of temperature and expected utility, represent non-extensive measurement.

To a *combination* structure $(M, \equiv, \leqslant, \circ)$, where M is some non-empty set, '$\equiv$' and '$\leqslant$' are relations between elements in M, and '$\circ$' is a two-termed operation for elements in M such that for all $x, y \in M$ it holds that $x \circ y \in M$, there corresponds either a combination scale $(P, =, <, +)$, where P is the set of positive real numbers, '$=$' denotes identity, '$<$' the relation *less than*, and

'+' the operation of addition, or an extended combination scale $(R =, <, +)$, where R is the set of all real numbers.

To an *interval* (difference) structure $(M, \equiv, \leqslant, /)$, where '/' is a four-termed relation between elements in M, there corresponds an interval (difference) scale $(R, =, <, /)$, where R is the set of all such real numbers for which $(x + y) / Z \in R$, when $x, y \in R$, and it holds for / that

$$/ (x, y, z, w) \leftrightarrow |x - y| \leqslant |z - w|.$$

To a *mixture* structure $(M, \equiv, \leqslant, [a])$, where $[a]$ is a two-termed operation in M for every a in the open interval $(0, 1)$, there corresponds a mixture scale $(R, =, <, [a])$, where R is the set of such real numbers for which $a \cdot x = (1 - a) \cdot y \in R$, when $x, y \in R$, and $[a]$ is defined for every a in $(0, 1)$ so that for $x, y \in R$

$$x [a] y = a \cdot x + (1 - a) \cdot y \text{ holds.}$$

It is interesting in this doctrine – though it obviously accentuates the chief ideas of the formal theory of measurement – that it admits the basic difference between fundamental and derived measurement, at least for physical measurement. Kanger regards derived measurement[12] as a *defined* measurement and discerns, apparently under the influence of Hempel, two cases: on the one hand, the given magnitude is defined with the help of other magnitudes, on the other hand, the scale values of the investigated magnitude are defined by means of measuring other magnitudes. However, we consider as much more important the fact that Kanger distinguishes between empirical and non-empirical structures and that he regards only the combination structure as an empirical one. If we compare the last elements in the above-mentioned structure and in the scales corresponding to them, the reason for this differentiation is quite evident. While, for the combination structures and both kinds of combination scales, the operation of concatenation differs from the operation of addition, we find in both other cases, in spite of different interpretations, the same sign, namely, / and $[a]$.

So when we think over thoroughly the basic presuppositions of the formal theory of measurement, the principal weaknesses of this theory will be revealed to us quite explicitly. If we cannot formally differentiate all components of structures and scales, and if we thus cannot find an appropriate empirical concept to every numerical concept, the structures, and likewise the scales, represent only abstract, idealized entities. Therefore, the stipulation of formal conditions, under which some structure would be homomorphic to a definite scale without, at the same time, having components which could be empirically interpreted and operationally realized, should not be viewed as a sufficient ground of measurability. Of course, this also holds the other way around: the definition of the concept of measurement cannot be given by a mere stipulation of operational procedures.

Both extreme positions – the formal theory and the operational doctrine of measurement – thus agree, in spite of their contrary starting points, in holding every measurement to be fundamental. The adherents of the formal theory of measurement stress direct numerical mapping of some empirical relational system, for they assume that this will help them to defend the broadest conception of measurement and to justify the autonomy of extraphysical measurement. To an operationalist, a derived measurement is not a measurement, because it rests on calculations, instead of directly performed measurement procedures.

The theory which has been elaborated by Hempel in connection with the analysis of the problems of concept formation in the empirical sciences has attempted to remove the most obvious shortcomings of the operationalist standpoint. In this analysis, Hempel strives to find more appropriate proportions between the empirico-operational and conceptually-mathematical aspects of measurement and, at the same time, to synthesize Campbell's and Stevens' understanding of measurement, yet in his exposition of the mutual relationship between fundamental and derived measurement,[13] he does not arrive at unambiguous conclusions.

Although, in accordance with Campbell, Hempel also distinguishes two basic kinds of measurement, yet he understands them as ways of introducing scalar quantitative notions in the sense of a theoretically oriented classification of concepts into qualitative, comparative, and quantitative, or as two distinct ways of defining scales of measurement or metrical scales in the spirit of Stevens' point of view. According to Hempel's view, fundamental measurement, which does not presuppose any previous scales of measurement, consists of two steps: in the specification of some comparative notion which defines a non-metric ordering, the so-called *quasi-series*, and in metrization of this ordering by introducing the respective numerical values. Hempel then regards as a derived measurement a determination of some metric scales with the help of criteria which assume at least one previous scale of measurement.

This specific conception of fundamental and derived measurement, worked out independently of the metrization of qualitative concepts, signifies, of course, an important shift in the interpretation of the concept of measurement. Yet this is demonstrated in a quite explicit form in the work of other writers. Thus, for example, Stegmüller[14] speaks only about the introduction of some metric magnitudes by means of a fundamental or a derived metrization and not, however, about its fundamental or derived measurement, for measurement is for him only an empirical procedure – a summary of measurement techniques and procedures – applied in the experimental verification of metrization. Fundamental (primary) measurement is conceived as the construction of some metric concept. Then he understands a derived (secondary) metrization, in the intentions of a neopositivistic reductionism, as a translation of some qualitative concepts into other metric concepts.

Hempel himself relates metrization, on the one hand, to the constitution of quantitative notions or metric scales, and on the other hand, to a fundamental measurement. In the spirit of Campbell's doctrine, he requires that the condition of additivity be satisfied for the fundamental measurement in physics and, therefore, regards every fundamental measurement of metric magnitudes as an instance of extensive measurement. At the same time, under the influence of Stevens, he concedes that there also exist measurement procedures, applied, for instance, in measuring the pitch of the tone in psychology, which are characteristic of fundamental measurement (they do not presuppose any previous scale of measurement), without, however, meeting the condition of additivity. Of course, Hempel's views are indebted to the operationalist doctrine of measurement, too, when he understands fundamental measurement as the actually performed measurement procedures, with the help of which the numerical values of the measured magnitudes are directly obtained. At the same time, Hempel is, however, very clearly aware that fundamental measurement procedures also depend on various conditions and presuppositions of a theoretical nature.

By a fundamental measurement, one can obtain only rational-number values, and then only under definite, operationally restricted conditions. If we measure, for example, mass by a fundamental measurement – more precisely, by means of directly applied measurement procedures – we might arrive at rational-number values only if we can weigh the measured objects. Yet, undoubtedly, we cannot measure in this way the mass of an entire range of empirical objects, for example, of the Sun or of an electron.

On the basis of this interpretation of the concept of fundamental measurement, taken under the influence of operationalism in the same sense in which one may understand the procedures of the so-called direct measurement, Hempel defines another essential characteristic for this way of measurement, the so-called *condition of commensurability*, which might be formulated as follows:

Every empirical object is commensurable with some standard object (measuring device) on the basis of the empirical relation of equality and the empirical relation of linking together (combining, chaining, concatenation).

With respect to the limited discriminatory abilities of an observer, one might, however, exclude the possibility of violating this condition in the fundamental measurement of physical magnitudes. Yet, as has already been mentioned, this additional condition of commensurability, which satisfies only the operationalist doctrine of measurement, is apparently unacceptable for theoretical reasons. For Hempel, this is also the principal argument against the operational defining of theoretical concepts. From the condition on commensurability it follows out of necessity that by a fundamental measurement – as far as he understands it as a collection of measurement procedures – one cannot obtain all numerical values which might be theoretically attributed to a given magnitude. Should we restrict the extension of the concept of fundamental measurement by this condition, it would necessarily mean that we would have to exclude from the interval of possible scale values not only all irrational values but also rational values, if the measured magnitudes are not immediately comparable with some standard object within the given range.

In order to bridge this conflict between the operational and theoretical aspects of fundamental measurement, Hempel seeks the way out through a modified version of the notion of derived measurement. This is quite inevitable for him if he wants to show that quantitative concepts in the empirical sciences are being defined by the theoretically founded rules of metrization. According to his view, one ought to distinguish between a so-called derived measurement (derived metrization) *by stipulation* and a derived measurement by means of *laws.*[16]

In the first case, some new metric concept is defined on the basis of notions which have already been introduced previously, for example, the average velocity of a point in a definite time interval as a ratio of distance and length of the time interval. In the second case, no new magnitude is introduced, but an alternative way of measurement that has not yet been used. This is possible only with the help of some law in which the given magnitude is represented as a mathematical function of other magnitudes, for which we already know suitable measurement procedures, for instance, the measurement of distance by sonar or radar, the measurement of altitude by a barometer, the measurement of temperature by a gas thermometer, and the like.

There exist various connections between fundamental measurement and both kinds of derived measurement, just as between a derived metrization by stipulation and a derived metrization by means of laws. By combining fundamental measurement with derived measurement by means of laws, we may establish numerical values of the measured magnitude that are not immediately accessible. If we combine, for instance, a fundamental measurement of length with a derived measurement on the basis of the laws of Euclidean geometry, we obtain irrational values, too. One might demonstrate the mutual relationship between these two sorts of derived measurement by giving the example of measuring temperature. The scale for temperature measurement was originally defined by stipulation with regard to the mercury thermometer. In this case, the notion of temperature can be interpreted only for materials within the range of their values between the point of thawing and the boiling point of mercury. Since an empirical law exists, according to which one may view temperature (under a constant pressure) in this range as a function of volume, we can measure the temperature of materials indirectly, on the ground of laws, by means of a gas thermometer, i.e., by virtue of establishing the volume which will be occupied by a definite standard amount of gas (under the specified pressure) in contact with the measured material. Should we now use a derived measurement by stipulation, we might decide that temperature will also exist outside the range of the original measurement of the functions of volume. Within the framework of measuring temperature by means of a mercury thermometer, the use of a gas thermometer represents a derived measurement on the basis of laws, while outside this framework it represents a derived measurement by means of stipulation.

From these relations between fundamental and derived measurement, as understood by Hempel, one cannot infer, however, that one and the same magnitude in a definite system of metric magnitudes would be measurable fundamentally at one time, while, at another time, derivatively, according to whether it is measured immediately (for example, if we measure the length of the side of a square), or with respect to certain laws (for example, if we measure the diagonal of the square).[17] In our view, this doctrine leads to an unwarranted switch between the notions of measurement and measurement procedures. However, it must be acknowledged that Hempel's definition of the concept of a derived measurement by means of laws refers, in fact, to the measurement procedures and, consequently, characterizes rather the so-called indirect measurement.

The difference between *direct* and *indirect* measurement concerns measurement procedures.[18] Direct measurement is based on an immediate comparison of the measured object with some standard object (measuring device) or with the scale of the measuring instrument. Indirect measurement includes a direct measurement of something else, of the same or other magnitudes, as well as calculations carried out on the basis of geometrical, physical, and other laws. In actual, (practically and theoretically relevant) cases, indirect measurements prevail, for one cannot measure directly the irrational values of magnitudes or the rational scale values, insofar as they refer to empirically and directly inaccessible values.

Direct and indirect measurement procedures can be applied to both derived and fundamental measurements. We can measure derivatively and directly the density of a homogeneous body by ascertaining its mass and volume. Indirectly, we can measure it on the basis of Archimedes' law, when instead of determining the volume, we weigh the body in a liquid. The distance between two accessible places is measured fundamentally and directly, while the distance up to some inaccessible place is measured indirectly, when, for instance, we utilize the measurement of angles and calculations performed with the help of geometrical laws.

The differentiation between derived measurement by stipulation and derived measurement by means of laws, implies the possibility of an additional meaning shift. Since, in an extraphysical measurement (if we do not want to put it into a relationship with physical measurement) we may hardly introduce some derived magnitudes (for the reasons which we have already specified), the concept of a derived measurement by stipulation is used to justify a quite arbitrary way of measurement, the so-called *measurement by means of convention* (*by fiat*).

We encounter this doctrine, for example, in Torgerson[19] who rests his trichotomic classification of the kinds of measurement (fundamental, derived, *by fiat*) on the so-called constitutive and operational meaning of the basic characteristics of the scales of measurement, namely on a theoretically justified interpretation of the conditions of ordering, distance function, and the origin of the scale, as well as on their operational realizability. In the case of a derived measurement that is based on laws, on objectively existing relations between various magnitudes or properties, the constitutive meaning of these characteristics is direct, yet their operational meaning is indirect. In the case of measurement "by fiat", which depends "on the assumed relation between observation and our interests", we may speak, at most, about their operational meaning. Of course, it is conceded by this that measurements carried out in this way – Torgerson mentions measurements in sociometrics, Pfanzagl–Bauman–Huber mention intelligence tests and measurements of social status – are theoretically unjustified. Since the measurement on the basis of conventions depends on the "intuition of each individual experimenter", the results obtained via this pseudo-measurement are considerably controversial. Although Torgerson strives to vindicate this sort of measurement by viewing it as practically important, he is compelled to admit that, due to the arbitrariness of the most diversified constructions of scales, some of which are entirely trivial, its trustworthiness is very low.

This negative consequence of Hempel's conception of the derived measurement by stipulation is compensated by Ellis' notion[20] of the so-called *associative measurement* which is linked with the concept of a derived measurement by means of laws. The immediate stimulus for the introduction of this kind of measurement is, as in Hempel, the problem area of measuring temperature. If we measure this quasi-extensive magnitude, which differs from the other metric magnitudes in physics due to its specific character, it will suffice to measure fundamentally or derivatively the length of some material used in thermometers, pressure, volume, electrical resistance, or any other thermometric magnitude, and to investigate the objective dependencies that exist between these magnitudes and temperature.

Then any kind of measurement in which some independently – fundamentally or derivatively – measured magnitude is tied up by functional relations with some quasi-extensive magnitude or qualitative property, is to be generally regarded as an associative measurement. By virtue of this mutual correlation, where every change of a directly measurable magnitude conditions a certain change of the magnitude or property which is associated with it, one can infer relevant conclusions concerning the scale values and numerical characteristics ascribed to magnitudes or properties that are not directly measurable. The associative measurement also differs from a derived one owing to the fact that it need not be based on numerical laws. We can also apply it in the case when definite qualitative laws exist, not only between non-metric and metric magnitudes, but also between non-metric magnitudes and quantitative data obtained by counting. Naturally, this is of great significance for extraphysical measurement.

After the acceptance of the associative measurement, Campbell's original dichotomic classification is extended into a trichotomic one in which fundamental, derived, and associative measurements then represent three basic kinds of measurement in the strict sense of the term. Therefore, we do not include among them so-called elemental measurement, illustrated by Ellis through the case of measuring the hardness of materials on Mohs' scale, for this sort of measurement satisfies only the conditions of topologization.

As has already been emphasized in the previous exposition, outside this classification of the *kinds of measurement* one has to distinguish, on the one hand, so-called direct and indirect measurement, and, on the other hand, so-called extensive and non-extensive (for example, interval) measurement. So-called direct and indirect measurement refers to the division of *measurement procedures*. The differentiation between so-called extensive and non-extensive measurement concerns the conceptual characterization of different *kinds of magnitudes* (metric, eventually also non-metric) and, consequently, also the manner of their metrization (or, perhaps, only topologization) and the resulting sort of scales.

6.3. METRIZATION

In the doctrine of Hempel and his followers, metrization, which comprises topologization as its necessary precondition refers above all to the constitution of quantitative (metric) concepts. The transition from classificatory notions to comparative ones, which fall under the class of qualitative concepts,[21] is theoretically defined by the satisfaction of topological conditions that determine the presuppositions under which some set of elements can be ordered in respect to some property. It is therefore the aim of topologization to find out whether the elements of the given set constitute a certain progression which is called *quasi-ordering* or *quasi-series*. Then the task of metrization of a qualitative concept, i.e., of the constitution of some quantitative (metric) concept, consists of finding a numerical function which would map the quasi-series, defined by topologization, onto the set of numbers. In a generalized formulation, when we think not only about a metrization of some qualitative concept, for example, the comparative concept *to be at least as hard as*, but also about a metrization of empirical relational systems, one understands under the concept of metrization[22] the establishment of a definite *metrization function* which would facilitate their homomorphic mapping onto a numerical system – onto the set of real numbers or onto the set of points of n-dimensional space. Besides this view, which is already influenced to a considerable extent by the formal theory of measurement, metrization of a quasi-series is defined, with a greater dependence on the initial intentions of this doctrine, by *metrization rules* and *measure principles.*[23]

What can be accepted from this position and what is to be rejected? If we understand measurement as a theoretically founded empirico-mathematical method (Section 2.3), we cannot one-sidedly emphasize only one of its components. We cannot talk merely about measurement or only about metrization and overlook, at the same time, their intrinsic connections. If we acknowledge that topologization is a conceptual method while metrization is a conceptually mathematical one, then we cannot view them as procedures of the same sort and level. Nevertheless, if we compare, for instance, Kutschera's understanding of metrization with Suppes' doctrine of measurement, or with Stevens' conception of the scales of measurement, with the exception of terminological distinctions, we shall not find here any essential difference in content. Considering the fact that many writers use the concept of metrization in the same sense as the concept of measurement and that every concept, which can be numerically represented inasmuch as it satisfies the conditions of topologization, is held to be a quantitative concept, scaling too falls into the extension of the concept of measurement. Yet one cannot agree with this consequence.

We assume, therefore, that it is necessary to re-interpret the notions of topologization and metrization so that they refer only to the conceptual aspects of measurement. Then, whether we consider the measurability of different kinds of magnitudes or only the measurement of some definite magnitude, topologization and metrization define, in a general or concrete form, the conditions that are to be satisfied by every empirical relational system, if it can be mapped onto a numerical relational system with certain formal characteristics and if, at the same time, we are to derive empirically meaningful conclusions as to the analogical properties of magnitudes.

This mapping might be expressed by *correspondence rules* which bring into correlation empirical and numerical characteristics of both relational systems, namely relations and operations defining their structure and, consequently, also the quantitative or only qualitative aspects of the measured magnitudes and scale values corresponding to them. The correspondence rules also define, therefore, the ways of interpreting the assigned scale values – whether they should be interpreted as cardinal numbers or only as ordinal numbers, hence, whether the scale values represent size too, or only an ordering. Evidently, their application is also limited by various idealizing conditions.

By means of topologization, one determines the conditions which must be satisfied by every non-metric magnitude. Yet, if topologization is not followed by a metrization, it constitutes only the conceptual basis of scaling. Metrization, which extends topological conditions by adding further assumptions that must be satisfied by metric magnitudes, is the conceptual basis of measurement. This differentiation between topological and metric conditions, and especially the limitation of the extent of metrization, should not, however, be understood so that we could not extend the scope of measurability also to comparative, or perhaps even to classificatory concepts, with the help of associative measurement. However, this possibility has not yet been sufficiently evaluated, theoretically researched, and practically utilized.

Under the influence of the formal theory of measurement, the attention of the methodologists of measurement has so far been particularly concentrated on the axiomatic stipulation of the formal properties of both these methods. In a whole series of axiomatizations,[24] which follow Hölder's system, these characteristics explicitly refer only to the systems of strictly extensive or extensive magnitudes and are therefore immediately tied up only with so-called extensive measurement. Only in connection with the most diversified extensions in the domains of social sciences, have topological and metric conditions also been axiomatically formulated for the various systems of so-called non-extensive (intensive) measurement. The representatives of extraphysical measurement[25] strive to apply them in various sorts of intervals or difference measurements, in the so-called bisection measurement (measurement by means of constructing the midpoint), in different (additive, multiplicative, and exponential) forms of conjoint measurement, or only in so-called ordinal measurement. As a consequence of this extension, metrization thus also refers – however unwarrantedly we may think – to the scaling of different non-metric magnitudes.

Since topological conditions are applicable to scaling, too, one may express topologization for the systems of non-extensive and extensive measurement, non-metric and metric magnitudes, in an analogical way.

Topologization, in its basic form, is defined by two primitive concepts: by binary relations K and P, or, alternatively, by binary relations K and L, defined for elements of some set $\boldsymbol{E}$. These relations might be generally interpreted as *coincidence* and *precedence*, or *coincidence* and *succession*. If we consider, for instance, the conceptual presuppositions of length measurement, we shall interpret K as the relation *to be equally long as*, P as the relation *to be shorter than*, and L as the relation *to be longer than*.

In order that the primitive relations K and P enable the ordering of elements $x, y, z, \ldots$ of the set $\boldsymbol{E}$, or more precisely speaking, their *weak ordering* of which it is assumed that it characterizes the majority of kinds of an empirical ordering, K and P must satisfy a whole range of properties.[26]

The relation K, namely, *x coincides with y*, must be *symmetrical*.

$$\text{(Ax)} \qquad (\forall x)(\forall y)(xKy \rightarrow yKx). \qquad (6.3\text{–}1)$$

Furthermore, the relation K must be *transitive*.

(Ax) $(\forall x)(\forall y)(\forall z)(xKy \wedge yKz \rightarrow xKz)$. (6.3–2)

Since from the symmetricity and transitivity, the property of reflexivity generally follows, the relation K must also be *reflexive.*[27]

(Th) $(\forall x)(xKx)$. (6.3–3)

The relation K is therefore a relation of the *equivalence* type. Owing to this relation, the set E is exhaustively divided into mutually exclusive equivalence classes of objects.

The relation P, namely, x *precedes* y, must be *transitive* too:

(Ax) $(\forall x)(\forall y)(\forall z)(xPy \wedge yPz \rightarrow xPz)$. (6.3–4)

In addition, the relation P must be *irreflexive*. Yet, in the given case, this is not enough. For a weak ordering of the set E by means of the relation P in respect to the relation K, it is also to be assumed that no element from E, for which it holds that it coincides with some other element, may be in relation P with this element. For this reason we require that the relation P be *K-irreflexive* (i.e., irreflexive with respect to K).

(Ax) $(\forall x)(\forall y)(xKy \rightarrow \neg(xPy))$. (6.3–5)

The relations P and K are mutually incompatible. All elements of the set E are to be comparable with respect to them. It must therefore be postulated that the relation P is connected as to the relation K, i.e., *K-connected*.

(Ax) $(\forall x)(\forall y)(\neg(xKy) \rightarrow xPy \vee yPx)$. (6.3–6)

This axiom can also expressed in the form

(Ax′) $(\forall x)(\forall y)(xKy \vee xPy \vee yPx)$, (6.3–7)

or, if we introduce the relation of *succession* L by the definition

(Df) xLy iff yPx, (6.3–8)

in the form

(Ax″) $(\forall x)(\forall y)(xKy \vee xPy \vee xLy)$. (6.3–9)

In this form it expresses the *law of trichotomy*. Nevertheless, we can interpret it only as a *weak* law of trichotomy, according to which, for all x and y, one of these alternatives holds, and not as a *strong* law of trichotomy, according to which, for all x and y, *exactly only* one of these alternatives holds. Such an interpretation would be in conflict with the assumption that topologization defines only a *quasi-series* that is isomorphic with the calculus of weak ordering.

This mutual relation is exhibited more distinctly if we interpret the set $\boldsymbol{E}$ as a set of numbers $a, b, c, \ldots$, P as the numerical relation '$<$' (or L as the numerical relation '$>$') and K as the numerical relation '$=$'. Then the above-mentioned axioms might be expressed in the form:

(Ax*)	$(\forall a)(\forall b)(a = b \to b = a)$	(6.3–1*)
(Ax*)	$(\forall a)(\forall b)(\forall c)(a = b \wedge b = c \to a = c)$	(6.3–2*)
(Ax*)	$(\forall a)(\forall b)(\forall c)(a < b \wedge b < c \to a < c)$	(6.3–4*)
(Ax*)	$(\forall a)(\forall b)(a = b \to a \nless b)$	(6.3–5*)
(Ax*)	$(\forall a)(\forall b)(a \neq b \to a < b \vee b < a)$.	(6.3–6*)

With the help of this interpretation we can then simplify the characterization of the conditions of topologization. For the relations between numbers $a, b, c, \ldots$, there also hold, as is known, the relations

$$a \leqq b \qquad \text{or} \qquad a \geqq b,$$

which we read, *a is smaller or equal to b (a is at most equal to b*), and, *a is greater or equal to b (a is at least equal to b*), respectively. The relation *smaller or equal* may be introduced by means of a definition

(Df) $\quad a \leqq b \quad$ iff $\quad a < b \vee a = b$. $\qquad$ (6.3–10*)

On the basis of this definition and the above-mentioned axioms, which have been interpreted by us as axioms of topologization as well as axioms of the calculus of weak ordering, we can derive the following two formulas in the numerical model

$$(\forall a)(\forall b)(a \leqq b \vee b \leqq a) \qquad (6.3\text{–}11^*)$$
$$(\forall a)(\forall b)(\forall c)(a \leqq b \wedge b \leqq c \to a \leqq c). \qquad (6.3\text{–}12^*)$$

The former expresses the condition of *connexivity*, while the latter expresses the condition of *transitivity*. If some numerical structure $\langle N, \leqq \rangle$ satisfies these two formulas, which might be regarded as variant axioms of the calculus of weak ordering (with the primitive notion '$\leqq$'), then every structure which is isomorphic to it can be weakly ordered.

If, on the contrary, we define the relation '$<$' and '$=$' on the basis of the primitive notion '$\leqq$'

(Df) $\quad a < b \quad$ iff $\quad \neg(b \leqq a)$ $\qquad$ (6.3–13*)

(Df) $\quad a = b \quad$ iff $\quad a \leqq b \wedge b \leqq a$. $\qquad$ (6.3–14*) from the axioms (6.3–11*) and (6.3–12*) we can derive as theorems the axioms (6.3–1*), (6.3–2*), (6.3–4*) – (6.3–6*). For the calculus of weak ordering we have thus at our disposal two equivalent axiomatizations: one is based on the primitive notion '$\leqq$' and two axioms, the other on the primitive notions '<' and '=' and five axioms.

This knowledge may be utilized for an alternative formulation of the conditions of topologization. If we interpret the relation '$\leqq$' as a binary relation Q in the sense, *x at most coincides with y*, and if we accept it as a primitive notion, then we can analogically formulate for topologization only two axioms, namely

$$\text{(Ax)} \quad (\forall x)(\forall y)(xQy \vee yQx) \tag{6.3–11}$$

$$\text{(Ax)} \quad (\forall x)(\forall y)(\forall z)(xQy \wedge yQz \rightarrow xQz). \tag{6.3–12}$$

However, we might likewise assign to the relation '$\geqq$' a binary relation Q in the sense, *x at least coincides with y*, to the relation '>' a binary relation L in the sense, *x succeeds y*, and to the relation '=' the relation K. The conditions of topologization may therefore be formulated either for the structures $\langle E, K, P\rangle$ and $\langle E, K, L\rangle$, or for the structure $\langle E, Q\rangle$. The second variant is theoretically "more elegant", for it is more concise. Methodologically, of course, the first – the original – variant gives a greater advantage, because the interpretation of the binary relation Q is not as intuitive as the one given for the relations K and P, or K and L. This holds even more for the operational realization of these relations.

From the one-to-one relation between the structures $\langle N, =, <\rangle$ and $\langle E, K, P\rangle$ of numerical and empirical relational systems, which are isomorphic to the calculus of weak ordering, there immediately follow the pertinent, generally formulated *correspondence rules*

$$\text{(Ru)} \quad a = b \Longleftrightarrow xKy \tag{6.3–15}$$

$$\text{(Ru)} \quad a < b \Longleftrightarrow xPy. \tag{6.3–16}$$

If we start from an alternative formulation of both relational systems, i.e., from the structures $\langle N, \leqq\rangle$ and $\langle E, Q\rangle$, it will suffice to give only one correspondence rule

$$\text{(Ru)} \quad a \leqq b \Longleftrightarrow xQy. \tag{6.3–17}$$

It is quite evident from the formulations of these rules that to arbitrary objects x, y, z of the set E, which coincide in respect to some non-metric or metric magnitude, we shall assign the same numerical values; if, however, in the given ordering x precedes y, we shall assign to the object x a smaller scale value than to the object y, or, alternatively, if x at most coincides with y, then we shall assign to the object x a smaller number than to the object y or the same number to both.

By means of the *correspondence rules of topologization*, formulated for the ordering of individual elements of some empirical set, one may therefore assign to different magnitudes only *ordinal scale values*.

The correspondence rules (6.3–15) and (6.3–16), or (6.3–17), must, however, be specified for each specific case. If we scale, for instance, the hardness of minerals, the first two rules may be expressed as follows:

(1) Only when the mineral n_x will not make an indentation in the mineral n_y, and vice-versa, the mineral n_y will not make an indentation in the mineral n_x, shall we assign to both minerals the same scale value or, alternatively, we shall assign the same numerical value to two minerals n_x and n_y precisely when the mineral n_x will not make an indentation into the mineral n_y, and vice-versa.

(2) Only when the mineral n_y makes an indentation into the mineral n_x shall we assign to the mineral n_x a smaller scale value than to the mineral n_y, or, alternatively, we shall assign a smaller numerical value to the mineral n_x than to the mineral n_y only when n_y makes an indentation into n_x.

If we measure, for example, the mass of bodies, we might interpret K as the relation *to be in an equilibrium* and P as the relation *to be lighter than*, and the like.

Since the correspondence rules bring into a mutual connection idealized mathematical entities and relations between them with concrete empirical objects and relations, it is obvious that they will be binding only approximatively, in spite of the explicitly expressed relation of one-to-one correlation. Their concrete application depends, for instance, on our ability to discern the relations of coincidence and precedence in practice, as well as on the possibility of finding for them some operationally realizable interpretation. Besides, we must not forget an additional fact that these relations cannot have exactly the same properties as the corresponding relations between numbers. It might hardly be assumed of the relation of coincidence in particular (but the same also holds with respect to the relation of precedence) that its transitivity would be practically testable for a large number of elements; whereas this property holds for numbers without any restriction. Also, the condition of comparability is to be viewed as an idealized supposition without which we could not accept the validity of the law of weak trichotomy. However, in practice we also encounter an incomparability of elements of the set E as to the relations K and P, or at least controversial cases, when we simply must make a decision in favor of one alternative, yet without being certain that this decision is the right one.

Still, in a much greater measure, we meet these problems in other kinds of topologization, specially proposed for the purposes of extraphysical measurement. A common mark of these topologizations is the fact that they use, besides the binary relations K and P, or Q, in addition to them or in a parallel way, still other quaternary or ternary relations with similar characteristics.

Metrization, as it has been worked out in connection with the analysis of the systems of extensive measurement, is then extending the conditions of topologization formulated for the primitive notions K, P, or K, L, or Q, and defined on the elements of the set E, by imposing further requirements referring to the operation S. This operation – $(xSy)Kz$ – in the sense of *x connected (combined, concatenated) with y coincides with z*, has been introduced as an empirical counterpart of the numerical operation of addition. Therefore, we lay down similar conditions for the operation S like the ones to be satisfied by the operation of numerical addition.[28] If we start from the empirical structure $\langle E, K, P, S\rangle$, we can define it as follows:

The operation S must be *commutative* and *associative*:

(Ax) $(\forall x)(\forall y)((xSy)\,K\,(ySx))$ (6.2–18)

(Ax) $(\forall x)(\forall y)(\forall z)((xS(ySz))\,K\,((xSy)Sz))$. (6.3–19)

The operation S must satisfy the so-called laws of *monotonicity* as to the relations K and P:

(Ax) $(\forall x)(\forall y)(\forall z)(\forall w)((xKy \wedge zKw) \rightarrow$
$\rightarrow (xSz)\,K\,(ySw))$ (6.3–20)

(Ax) $(\forall x)(\forall y)(\forall z)(\forall w)((xKy \wedge zPw) \rightarrow$
$\rightarrow (xSz)\,P\,(ySw))$ (6.3–21)

(Ax) $(\forall x)(\forall y)(\forall z)(\forall w)((xPy \wedge zPw) \rightarrow$
$\rightarrow (xSz)\,P\,(ySw))$. (6.3–22)

For the structure $\langle E, Q\rangle$ the laws of monotonicity may be expressed in a simpler manner:

$$(\forall x)(\forall y)(\forall z)(xQy \rightarrow (xSz)\,Q\,(ySz)). \qquad (6.3\text{–}23)$$

Because one can always introduce the inverse operation of subtraction to the numerical operation of addition, we must postulate for S too an inverse operation of *difference*

(Ax) $(\forall x)(\forall y)(\exists z)(xPy \rightarrow yK(xSz))$. (6.3–24)

In analogy with the axiomatic system of arithmetic that is limited to positive integers, yet another condition is introduced, according to which the existence of the so-called *zero element* in the set E is excluded

$$\text{(Ax)} \qquad (\forall x)(\forall y)(\forall z)(xKy \rightarrow xP(ySz)). \qquad (6.3\text{–}25)$$

If we take it consistently, we would have to postulate, in addition, the condition

$$(\forall z)(\forall y)(\exists n)(xPy \rightarrow yPnx), \qquad (6.3\text{–}26)$$

where $n = 1, 2, 3, \ldots$, and nx is recursively defined as follows: $1x = x$, and $(m + 1)x = m(xSz)$. This condition would correspond to the Archimedean law, familiar from the axiomatization of arithmetic. Nevertheless, no empirical set of actually measured objects, which we in fact measure, is practically infinite. In the same way, one cannot require that this set be closed with respect to the operation S, i.e., that it should hold for all x and y which belong to E that also their connection xSy belong to this set. Hence, the structure $\langle E, K, P, S\rangle$ of the empirical relational systems of extensive measurement is not – strictly speaking – isomorphic to the numerical structure $\langle N, =, <, +\rangle$, where N is the set of positive integers, either. Naturally, this holds even more for the mapping of some empirical relational system onto the real-number system.

Nevertheless, the above-mentioned conditions of metrization provide sufficient grounds for the formulation of another correspondence rule, in spite of the reservations and limitations that have already been discussed in our exposition of the conditions of commensurability and topologization. This rule is expressed in the following way:

$$\text{(Ru)} \qquad a + b = c \Longleftrightarrow (xSy)Kz. \qquad (6.3\text{–}27)$$

On the basis of this *correspondence rule of metrization* cardinal scale values can be assigned to various magnitudes. Its application assumes, therefore, that we are able to establish a measurement unit for the measured magnitude.

However, since quasi-extensive magnitudes (for example, temperature, though for it we have at our disposal a measurement unit) do not satisfy the condition of additivity postulated by this correspondence rule, we must modify the conditions of metrization in some way so that they would refer to all metric magnitudes.

In order to establish modified conditions of metrization which would characterize the measurement of temperature, one might utilize the knowledge which concerns associative measurement. Because there exist familiar functional relations between the changes of temperature and the expansion of materials used in thermometers, we can assign the difference of temperature to length distances, say, on the thermometer tubes. Then to a definite temperature difference there corresponds just one length distance

$$d(x_t, y_t) \quad \text{iff} \quad d(x_d, y_d).$$

The size of the length distance $d(x_d, y_d)$ may, at the same time, be chosen so that the value of the corresponding temperature interval $d(x_t, y_t)$ is equal to the measurement unit of temperature, its multiple or portion.

On the ground of the above-mentioned equivalence, the assigned scale value will be correlated analogically, in the given case: the scale distances

$$|a_t - b_t| \quad \text{iff} \quad |a_d - b_d|.$$

It is important for a progression of scale distances that, in both cases, we can represent them on the uniform scale as a progression of equally spaced intervals. This possibility follows from the fact that we have at our disposal measurement units for both these magnitudes.

Since length distances permit the application of all conditions of metrization and of the correspondence rules that are entailed by them, we can assume the same for the corresponding temperature differences, even though the requirement of additivity cannot be operationally realizable in their case. It follows from this justifiable presupposition that the correspondence rules for the relations K and P and the operation S, reflecting the essence of the metrization of metric magnitudes, can be reformulated for the measurement of temperature so that they will not refer to elements $x, y, z, \ldots$ from the set E and to numbers $a, b, c, d, \ldots$ from the set N, but to pairs of these elements – to temperature intervals and numerical differences. In a generalized form (without specification indices), they might be expressed as follows:

$$\text{(Ru)} \quad |a - b| = |c - d| \Longleftrightarrow d(x, y) K\, d(z, w) \qquad (6.3\text{–}15')$$

$$\text{(Ru)} \quad |a - b| < |c - d| \Longleftrightarrow d(x, y) P\, d(z, w) \qquad (6.3\text{–}16')$$

$$\text{(Ru)} \quad |a - b| + |b - c| = |a - c| \Longleftrightarrow (d(x, y) S\, d(y, z)) K\, d(x, z). \qquad (6.3\text{–}27')$$

Apparently, the third rule is restricted by the condition that temperature intervals must be immediately connected with each other.

These *correspondence rules of topologization and metrization*, formulated for the pairs of elements from an empirical set, refer to every structure of an interval (difference) measurement for which one cannot explicitly introduce the operation of concatenation. Yet they can be applied to other magnitudes only when the presuppositions under which they had been introduced are satisfied.

We shall regard an interval measurement of magnitudes, for which one cannot establish a measurement unit and which cannot be associated with some extensively measurable magnitude, only as a scaling. It is this point of view from which one has to judge Pfanzagl's understanding of metric connection and distance function (Section 4.3), just like various topologizations and metrizations in the theories of extraphysical measurement.

The most elaborated version of the topologization of non-metric magnitudes, striving to interpret it at the same time as a metrization, is tied up with the analysis of different kinds of *interval* (*difference*) *measurement*. This sort of a non-extensive measurement is characterized by different empirical structures: for example, by the structure $\langle E, K, P, / \rangle$, where / is a quaternary relation (Section 6.2); by the structure $\langle E, E^x, D \rangle$, where E^x is a non-empty subset of the Cartesian product $E \times E$ and D is a binary relation on E^x, i.e., a two-termed relation between the pairs of elements from E; or by the structure $\langle E, \boldsymbol{D} \rangle$, where $\boldsymbol{D}$ is a quaternary relation

$$\boldsymbol{D}(x, y, z, w) \quad \text{or} \quad x, y \ \boldsymbol{D} \ z, w.$$

All these structures are alike as to the occurrence of some quaternary relation, which can be interpreted from a formal point of view like the relation Q or the inverse relation $\breve{Q}$. Their difference depends on the way the following two questions are answered: (a) What are the empirical differences to which this relation is being assigned? (b) Is it assumed about the empirical structure that it is finite, or is it assumed that it is infinite? Yet these differences are inessential when it comes to apprehending the core of these problems.

The relation $\boldsymbol{D}$ holds between the pairs of elements x, y, and z, w from the set E, if we can assign to them numbers a, b, c, d so that the numerical difference between a and b is at least equal (greater or equal) to the numerical difference between c and d. It is usually assumed about these numerical differences that their value is positive; yet, a contrary case, of course, is not excluded. If we compare, for instance, different pairs of elements as to their similarity, say, with regard to their color or shape, we can interpret the relation $\boldsymbol{D}$ as follows: *the similarity between x and y is at least so great* (*greater or equal*) *as the similarity between z and w*.

In a general interpretation, these relations are most frequently explicated as the relations *to be at most indifferent as* or *to be at least as much indifferent as* and *to be at most* (*at least*) *as much indiscernible as*. From the formal and interpretational points of view, the distinction between them is irrelevant, if we disregard here the way of understanding the notion of an interval (as absolute, positive, or some other). If we consider the intended numerical interpretation, for which it holds

$$a - b \lesseqqgtr c - d \longleftrightarrow c - d \gtreqqless a - b, \tag{6.3–28*}$$

then we can analogically presuppose the validity of the equivalence: x, y is, at most, as indifferent as z, w iff z, w is at least as indifferent as x, y.

The best illustration of the characterization of the conditions of topologization for various kinds of interval measurements, differentiated according to the cardinality of the set E and the possible interpretations of the concept of a numerical interval, might be provided by the analysis of a *finite, equally spaced interval structure* $\langle E, \boldsymbol{D}\rangle$.

By means of the primitive notion $\boldsymbol{D}$, which has the intended numerical interpretation '$\geqq$', one may analogically define, as with the help of the relation Q (or the inverse relation $\breve{Q}$), a series of other quaternary relational notions: the concept of preference $\boldsymbol{V}$[29]

$$\text{(Df)} \qquad x, y\ \boldsymbol{V}\ z, w \quad \text{iff} \quad \neg(z, w\ \boldsymbol{D}\ x, y) \tag{6.3–29}$$

and the concept of indifference (indiscernibility) $\boldsymbol{U}$

$$\text{(Df)} \qquad x, y\ \boldsymbol{U}\ z, w \quad \text{iff} \quad x, y\ \boldsymbol{D}\ z, w \wedge z, w\ \boldsymbol{D}\ x, y \tag{6.3–30}$$

Besides these quaternary relational notions, with the help of $\boldsymbol{D}$ we can furthermore define a binary relation of indifference (indiscernibility) U

$$\text{(Df)} \qquad x\,U\,y \quad \text{iff} \quad x, y\ \boldsymbol{U}\ y, x \tag{6.3–31}$$

and a binary relation of preference V

$$\text{(Df)} \qquad x\,V\,y \quad \text{iff} \quad x, x\ \boldsymbol{U}\ y, x. \tag{6.3–32}$$

Thus we can also define – either by means of the relations $\boldsymbol{U}$ and $\boldsymbol{V}$ or by means of the relation $\boldsymbol{D}$ – a binary relation of indifference or preference $U // V$

$$\text{(Df)} \qquad x\,U // V\,y \quad \text{iff} \quad x, x\ \boldsymbol{D}\ y, x. \tag{6.3–33}$$

It is, however, much more important for the needs of topologization of the empirical structure $\langle E, \boldsymbol{D}\rangle$, if it is to satisfy the condition of equally spaced intervals, to introduce another notion – a binary relation J, the relation of the so-called *immediate predecessor*

(Df) $x\, J\, y$ iff $x\, V\, y \wedge (\forall z)\,(z\, V\, z \to y\, U\, z \vee y\, V\, z)$. (6.3–34)

The intuitive sense of this definition is obvious: under the conditions stated in the definiens, all immediately adjacent elements in the given ordering are equally distant from each other. Therefore, if x and y are two such elements, no element z may occur between them. If instead of the relation $\boldsymbol{D}$ we would introduce as a primitive notion the relation / (with the intended interpretation '$\leqq$'), we would have to introduce, on the contrary, the notion of *immediate successor*, by means of which one might generate an equally spaced progression of these numbers, in accordance with Peano's axiomatization of natural numbers.

Nevertheless, the analogy between a progression of numbers and a series of equally spaced elements from the set E should not be understood so that the interval between arbitrary elements x and y would be equal to 1. If we express the definiens in the numerical form

$$a > b \wedge (\forall c)\,(a > c \to b = c \vee b > c),$$

this formula will be, in addition, satisfied, not only for $a = 8$, $b = 7$, $c = 6$, but also, for example, for $a = 8$, $b = 6$, $c = 2$. It will be unsatisfied only for every c such that $c > b$, for instance, for $a = 8$, $b = 4$, $c = 5$, because in this case c will be between a and b; thus the corresponding elements x and y will not be immediately adjacent elements of the given ordering.

It is evident, on the basis of these definitions, that for the topologization of an interval measurement, one might also give two equivalent variants with similar axioms like those for the structures $\langle E, Q\rangle$ and $\langle E, K, P\rangle$: one will be based on the structure $\langle E, \boldsymbol{D}\rangle$, the other on the structure $\langle E, \boldsymbol{U}, \boldsymbol{V}\rangle$. To simplify our exposition, it will suffice to offer only the former variant; for the sake of economy we shall omit, at the same time, universal quantifiers too:

(Ax) $x, y\ \boldsymbol{D}\ z, w \vee z, w\ \boldsymbol{D}\ x, y$ (6.3–35)

(Ax) $x, y\ \boldsymbol{D}\ z, w \wedge z, w\ \boldsymbol{D}\ u, v \to x, y\ \boldsymbol{D}\ u, v$. (6.3–36)

The first axiom characterizes the condition of *connexivity*, while the second characterizes the condition of *transitivity*, for the relation $\boldsymbol{D}$. Hence, this also expresses the fact that this quaternary relation allows a *weak ordering* of the pairs of elements from the set E. Because the notion of an empirical interval (difference, distance), which has been introduced in the theory of measurement of non-extensive magnitudes *per analogiam* as a qualitative correlate of the arithmetical concept of difference (interval), or of the geometrical concept of distance (segment), is not itself objectively founded to a sufficient degree, two additional axioms are introduced:

(Ax) $x, y \, \boldsymbol{D} \, z, w \longrightarrow x, z \, \boldsymbol{D} \, y, w$ (6.3–37)
(Ax) $x, y \, \boldsymbol{D} \, z, w \longrightarrow w, z \, \boldsymbol{D} \, y, x.$ (6.3–38)

The first axiom characterizes a necessary property of the intended numerical interpretation for the relation $\boldsymbol{D}$, and the second axiom – a necessary algebraic fact concerning the differences, namely

(Ax*) $a - b \geqq c - d \longrightarrow a - c \geqq b - d$ (6.3–37*)
(Ax*) $a - b \geqq c - d \longrightarrow d - c \geqq b - a.$ (6.3–38*)

Instead of the axiom (6.3–38) there is sometimes also given the axiom

$$x, y \, \boldsymbol{D} \, x', y' \wedge y, z \, \boldsymbol{D} \, y', z' \longrightarrow x, y \, \boldsymbol{D} \, x', y'. \qquad (6.3\text{–}39)$$

From these axioms, which are common to all structures of an extraphysical interval (difference) measurement, there immediately follows the *correspondence rule*

(Ru) $a - b \geqq c - d \Longleftrightarrow x, y \, \boldsymbol{D} \, z, w,$ (6.3–40)

which is equivalent to both first rules for the interval measurement of temperature, formulated for $d(x, y)$ and $d(z, w)$.

The last axiom expresses the specific presupposition of an *equal spacing*: if x is immediately adjacent to y and z is immediately adjacent to w, then the difference between x and y must be the same as the difference between z and w:

(Ax) $x \, J \, y \wedge z \, J \, w \longrightarrow x, y \, \boldsymbol{U} \, z, w.$ (6.3–41)

This axiom is used only at the formulation of the conditions under which the scale form of interval structures is invariant, or in the derivation of the respective uniqueness theorem. However, the assumption of equal spacing for non-metric magnitudes is, in fact, merely a fictitious surrogate for the non-existence of a measurement unit and therefore it does not authorize us to infer the same consequences as those which refer to the seemingly analogical case of the interval measurement of temperature.

In following the above-mentioned exposition of an interval measurement, we shall now approach a concise outline of the topologization of the system of so-called bisection measurement (the construction of midpoints), as well as the simplest versions of the additive conjoint measurement.

The measurement by means of *constructing the midpoints* through the so-called *bisection of intervals*, which was introduced in psychophysical measurement by G. T. Fechner during the last century, consists of the procedure in which the subject strives to choose some stimulus which he subjectively perceives, for example, as loudness, in such way that this stimulus lies exactly in the middle of two other stimuli. We might therefore view this measurement as a case of associative measurement. The methodologists of extraphysical measurement who strive to construct a theory of psychological measurement independent of physical measurement, subsume this kind of measurement either under the system of the so-called *bisymmetric* measurement or under the system of *conjoint* measurement.

From a structural viewpoint, however, we might also adequately characterize the measurement by means of constructing the midpoints by an empirical structure of the systems of an interval measurement $\langle E, \mathbf{D} \rangle$ with a binary relation J, if in addition we introduce two ternary relations: the relation $\mathbf{M}(x, y, z)$ having the meaning of *y lies between x and z*, and the relation $\mathbf{B}(x, y, z)$ which has the meaning of *y lies in the middle between x and z*:

$$\text{(Df)} \quad \mathbf{M}(x, y, z) \quad \text{iff} \quad xVy \wedge yVz \vee zVy \wedge yVx \qquad (6.3\text{–}42)$$

$$\text{(Df)} \quad \mathbf{B}(x, y, z) \quad \text{iff} \quad \mathbf{M}(x, y, z) \wedge x, y \; \mathbf{U} \; y, z. \qquad (6.3\text{–}43)$$

Yet this notion may also be immediately defined, with the help of the binary relation J

$$\text{(Df)} \quad \mathbf{B}(x, y, z) \quad \text{iff} \quad xJy \wedge yJz. \qquad (6.3\text{–}44)$$

The axioms in terms of which we may characterize the topological conditions for the measurement by means of constructing the midpoints, are intuitively evident. Individual elements of the set E are weakly ordered and their progression is gradually equally spaced so that, for arbitrary two boundary elements of an interval, one can always establish some third, midpoint element.

The corresponding numerical mapping within the *correspondence rules* that can be expressed by means of the relations U, V and $\mathbf{B}$, or, in an equivalent way, by means of the relations K, P (or L) and $\mathbf{B}$, if we want to point out the connection between these rules and the correspondence rules for the systems of extensive measurement, namely

$$\text{(Ru)} \quad a = b \Longleftrightarrow xKy \qquad (6.3\text{–}15)$$

$$\text{(Ru)} \quad a < b \Longleftrightarrow xPy \qquad (6.3\text{–}16)$$

$$\text{(Ru)} \quad (a + b)/2 \Longleftrightarrow \mathbf{B}(x, y, z), \qquad (6.3\text{–}45)$$

reminds us, indeed, rather of the metrization of the system of extensive measurement.

Yet this impression is entirely without foundation. Measurement by means of constructing midpoints, as it is understood today by the representatives of extraphysical measurement, is a specific case of a non-extensive measurement. Thus, for the measured magnitudes, one may not stipulate a measurement unit *ex definitione*. The assigned scale values provide only for the ordering of the measured magnitudes and not, however, also for their size. Even less can we find for them an empirically meaningful interpretation of the numerical operation of addition. The validity of the third correspondence rule is therefore justifiable only if we can assume that the non-metric magnitude, which we wish to measure in this manner, is associatively measurable.

For a *conjoint measurement*, which is grounded on the presupposition of a simultaneous measurement or scaling of at least two different magnitudes or properties (for example, intensity and frequency of tones), there is postulated an empirical structure $\langle E_1, E_2, \boldsymbol{D}' \rangle$, where E_1 and E_2 are two different empirical sets of objects. The quaternary relation $\boldsymbol{D}'$ – $\langle x, i \rangle\ \boldsymbol{D}' \langle y, j \rangle$ – refers to the ordered pairs $\langle x, i \rangle$ and $\langle y, j \rangle$, where x and y are elements of the set E_1 and i and j are elements of the set E_2. The relation $\boldsymbol{D}'$ differs from the analogical quaternary relation $\boldsymbol{D}$ of an interval measurement only by the fact that the Cartesian product $E_1 \times E_2$ is here defined for two distinct empirical sets and that the members of the relation $\boldsymbol{D}'$ are ordered pairs and not intervals or differences between elements of the set E.

With the help of the relation $\boldsymbol{D}'$ one might introduce analogically, according to the definitions (6.3–29) and (6.3–30), two quaternary relations $\boldsymbol{V}'$ and $\boldsymbol{U}'$, just as three binary relations U', V' and $U' \,//\, V'$ (compare definitions (6.3–31)–(6.3–33)). Now these binary relations, however, are to be defined with regard to the first and second components, for x, y and i, j belong to two different empirical sets,

$$\text{(Df)} \quad xU' \,//\, V'y \quad \text{iff} \quad (\forall i)(\langle x, i \rangle\ \boldsymbol{D}' \langle y, i \rangle) \qquad (6.3\text{–}46)$$

$$\text{(Df)} \quad iU' \,//\, V'j \quad \text{iff} \quad (\forall x)(\langle x, i \rangle\ \boldsymbol{D}' \langle x, j \rangle), \qquad (6.3\text{–}47)$$

and this, of course, holds for the similar definitions of the relations $\boldsymbol{U}'$ and $\boldsymbol{V}'$. In limiting ourselves to a finite, equally spaced additive conjoint measurement, we must assume, furthermore, a binary relation J.

The axiomatic formulation of the conditions of topologization then encompasses, besides the already familiar axioms of weak ordering (i.e., the axioms of connexivity and transitivity), two axioms of *independence*

(Ax) $\langle x, i\rangle\ \mathbf{D}'\ \langle y, i\rangle \longrightarrow \langle x, j\rangle\ \mathbf{D}'\ \langle y, j\rangle$ (6.3–48)
(Ax) $\langle x, i\rangle\ \mathbf{D}'\ \langle x, j\rangle \longrightarrow \langle y, i\rangle\ \mathbf{D}'\ \langle y, j\rangle$. (6.3–49)

The former can be interpreted as follows: if the ordered pair $\langle x, i\rangle$ is at least as much preferred as the ordered pair $\langle y, j\rangle$, then this relation also holds for the case of replacing the element i by an arbitrary other element j from the set E_2. The latter axiom assigns the same property to elements of the set E_2 with respect to elements of the set E_1.

Another axiom, the last one, postulates the condition of an *equal spacing*,

(Ax) $xJy \wedge iJj \longrightarrow \langle x, j\rangle\ \mathbf{U}'\ \langle y, i\rangle$, (6.3–50)

which again is to serve as a substitute for the non-existence of measurement units conjointly measured, yet, in fact, only conjointly scaled, non-metric magnitudes.

The structure of this basic variant of conjoint measurement satisfies, therefore, the conditions of weak ordering. Nevertheless, since it is axiomatically postulated as to the elements of the sets E_1 and E_2 that they are mutually independent, for each of these sets it will be necessary to formulate similar correspondence rules which contain in their numerical component the relations '=' and '>' (or '<'), or, equivalently, '≧' (or '≦'). Of course, the satisfaction of these common conditions of topologization is quite trivial and does not distinguish the structure of conjoint measurement from other forms of non-extensive measurement. Hence, for the equally spaced additive conjoint measurement there is introduced a specific *correspondence rule*

(Ru) $a + d = b + c \Longleftrightarrow \langle x, j\rangle\ \mathbf{U}'\ \langle y, j\rangle$. (6.3–51)

Then, according to this rule, to elements x, y from the empirical set E_1, as well as, under the same conditions, to elements i, j from the empirical set E_2, there are assigned arbitrary numbers a, b for which $a \geqq b$ holds. So far as there exists the same difference between the pairs of numbers a, b and c, d – postulated by the axiom of an equal spacing for the corresponding elements of the empirical sets – the expression $\langle x, j\rangle\ \mathbf{U}'\ \langle y, i\rangle$ will, of course, be meaningful precisely only when $a + d = b + c$ (for example, let $a = 3$ for x and let $b = 1$ for y; further, let $c = 9$ for i and let $d = 7$ for j).

The foregoing correspondence rule is also formulated more generally in the form

(Ru) $a + d \geqq b + c \Longleftrightarrow \langle x, j\rangle\ \mathbf{D}'\ \langle y, i\rangle$. (6.3–52)

Nevertheless, one cannot infer from this that this rule might also be applied to the relation V' and to its numerical counterpart '>', for this would be in conflict with the requirement of an equal spacing.

The above correspondence rule, in this or another formulation, immediately follows from the axiom of equal spacing and at least partially illuminates the intended interpretation of the concept of an ordered pair of the type $\langle x, i \rangle$. What, however, remains unexplainable is the problem of how to add the assigned scale values and, thus, how to combine elements of two empirical sets that are differentiated as to their content. Suppes' claim [30] that real-valued functions are combined together on each component with the help of the same unit, is unjustified for matter-of-fact reasons; at any rate, it also conflicts with the basic presuppositions from which the theory of conjoint measurement originates. If we had at our disposal measurement units for both these magnitudes – where, of course, the measurement units would have to be different – and if we were able to give for them empirically realizable conditions which would permit the combining of elements of both sets in accordance with the assumed additivity, then indeed it would not be necessary by means of very complicated constructions to introduce this specific sort of non-extensive measurement.

The postulation of different conditions of metrization (which, however, cannot be practically realized) for non-extensive magnitudes, too, is not a passable way that would lead toward spanning the objectively existing difference between what is measurable and what is not, just as between what is fundamentally measurable and what is not. Furthermore, the construction of purely formal systems, which in fact are appropriate only to some numerical model, is not even a sufficient ground for the justification of the possibility of measuring these magnitudes, at least associatively.

6.4. THE REPRESENTATION THEOREM

As we have already mentioned, the idea of formulating the problem of representation and of deriving some representation theorem for every axiomatically introduced empirical structure of measurement, had been directly inspired by the axiomatic theory of expected utility. In an attempt to indirectly justify the plausibility of Bernoulli's hypothesis of maximization of expected utility, von Neumann and Morgenstern [31] derive both basic properties of this notion, namely

$$u \gtrsim v \quad \text{implies} \quad \mathrm{v}(u) \geqq \mathrm{v}(v), \qquad (6.4\text{–}1)$$

in the sense "if an individual holds the alternative (utility) u to be at least as desirable as the alternative (utility) ν, he will assign to the utility function $\mathrm{v}(u)$ a greater or equal numerical value than to the utility function $\mathrm{v}(\nu)$", and

$$\mathrm{v}(\gamma u + (1 - \gamma)\nu) = \gamma \mathrm{v}(u) + (1 - \gamma)\mathrm{v}(\nu), \tag{6.4–2}$$

having the interpretation "the expected utility (utility function) of the combined probability of both alternatives (utilities) u and ν, i.e., of the mixed alternative $\gamma u + (1 - \gamma)\nu$, where $0 < \gamma < 1$, is identical (equal) to the expected utility (utility function) of their probabilistically weighed center of gravity" in the form of the so-called *existence theorem* (A : V)

"There exists a mapping

$$w \Rightarrow \mathrm{v}(w)$$

of all w on a set of numbers possessing the following properties:

(i) Monotony
(ii) For $0 < \gamma < 1$ and any u, ν, $\mathrm{v}((1 - \gamma)u + \gamma\nu) =$
$= (1 - \gamma)\mathrm{v}(u) + \gamma\mathrm{v}(\nu)$."

The problem area of representation, the positive solution of which ought to lead to a definite representation theorem, has been generally postulated as an inevitable ingredient of every axiomatic theory of measurement by Scott and Suppes (1958) and explicitly declared by Suppes and Zinnes (1963) to be one of the fundamental goals in studying the foundations of the theory of measurement.

The formulation of the representation theorem, as far as it immediately refers to a definite system of fundamentally measurable magnitudes, does not essentially differ from the formulation of the correspondence rules. At first glance, this may be seen in comparing, for instance, the correspondence rules for the systems of extensive measurement, namely (6.3–17) and (6.3–27), with the following formulation of the representation theorem:[32]

"Let A be a nonempty set, $\gtrsim$ a binary relation on A, and $\circ$ a closed binary operation on A. Then $\langle A, \gtrsim, \circ \rangle$ is a closed extensive structure iff there exists a real-valued function ϕ on A (ϕ: $A \to \mathbf{Re}$) such that for all $a, b \in A$

(i) $a \gtrsim b$ iff $\phi(a) \geqq \phi(b)$;
(ii) $\phi(a \circ b) = \phi(a) + \phi(b)$."

Hence, the representation theorem specifies, in a similar way to the correspondence rules, the conditions under which the given empirical relational system is homomorphic (or isomorphic) to some relational numerical system. Unless a distinction is drawn – especially in psychological literature or in the formal theory of measurement – between the concepts of measurement and scale, this is sometimes also stated by contending that in both cases conditions are defined under which some metric, or also non-metric, scales might be constructed.

The correspondence rules differ from the representation theorem owing to the fact that they correlate empirical and numerical concepts so that they can be applied only in the case of fundamentally measurable magnitudes. Yet one might be proving the representation theorems for the systems of derivatively measurable magnitudes, too.[33] Thus, for example, for the measurement of density, defined by the structure $\langle E, \phi_m, \phi_V \rangle$, where E is a set of empirical objects, ϕ_m is a function of mass, and ϕ_V is a function of volume, there exists a representation relation $Z(\phi_m, \phi_V, \psi)$ such that, for every $x \in E$

$$(x) = \frac{\phi_m(x)}{\phi_V(x)}.$$

Another difference might be seen in the fact that the correspondence rules for physically measured magnitudes, which start from the familiar properties of numerical relational systems, can be relatively easily formulated and empirically interpreted. The explication of these rules is intuitively evident, so much so that it is unnecessary to justify this deductively – as it is requested in the derivation of representation theorems from the axioms postulated for various empirical structures of the systems of non-extensive magnitudes. The correspondence rules directly specify for which numerical relations and operations we have to find appropriate empirical correlations possessing similar properties, facilitating the assignment of scale values to quantitative aspects of the measured objects, so that they would not only express their ordering, but also their size. Although the derivation of representation theorems follows, in principle, the same aims, it is nevertheless necessary – owing to the difficulties connected with the empirical interpretation of numerical concepts and with the operational realizability of the assumed conditions of metrization – to elucidate first the structure of some empirical relational system which could have similar properties to a numerical system.

The justification of representation theorems via axiomatization is seemingly more evident than the immediate formulations of correspondence rules. Yet, in fact, it only hides the problematic character of many proposed systems of extraphysical measurement that cannot be operationally realized in a satisfactory way. The proponents of the formal theory of measurement are aware of this principal problem, yet, from the standpoint of their orientation, they do not regard it as a shortcoming. In their view, it is not an objective of the axiomatic justification of representation theorems to find out whether "a certain procedure is possible, but rather that a numerical function satisfying certain properties exists."[34] Therefore, they take as legitimate only the consideration of abstract structures of measurement,[35] characterized by abstract laws, without assuming necessarily that these structure are also materialized in practice.

Naturally, such a purely formal grounding of the theory of measurement is not sufficient. A derivation of a representation theorem cannot be an aim in itself. Metrization makes sense only in connection with measurement, if it helps us to theoretically clarify measurements carried on in practice and if it is verified by such measurements. Hence, a proof of the existence of some numerical function, offered by means of a derivation of some representation theorem, does not yet mean the possibility of a real construction of this function. In deriving some representation theorem from a set of axioms, which is at least consistent, we can demonstrate only its *logical* existence, i.e., its consistency. However, if the axioms could not be empirically interpreted and verified in practice, we could not justify with their help that such a numerical function also exists *factually*. Yet, if we cannot give an appropriate empirical interpretation of the primitive notions of empirical structures – an interpretation which would meet the intended system of measurement – we cannot do so to the representation theorem either. The burden of the proof of the supposed correlation between numerical relations and operations and their empirical counterparts will thus only be transferred to a more abstract plane. Because no problem of content may be meritoriously solved by formal means, this approach contributes rather to making the entire problem area more obscure, and definitely not clearer.

We can make this evaluation more concrete by giving two specific characteristics of the problems of representation in the theories of extraphysical measurement, which in either case follow the same intention: to justify the possibility of a fundamental measurement for quasi-extensive and non-metric magnitudes, too.

Since the function ϕ is also generally held to be a real-valued function in the theories of extraphysical measurement, when formulating the problem of representation we can hardly overlook the question of what indeed corresponds to the numerical operation of addition. Obviously, this question is very unpleasant. So that it would not be necessary at all to set it forth and then to seek various excuses, Suppes and other authors utilize the possibility of formally describing every binary operation by means of some ternary relation.[36] Nevertheless, a formal description of an empirical structure, in which the operation of concatenation would be expressed with the help of a ternary relation on $\boldsymbol{E}$ holding between x, y and z $K(xSy)$, does not solve anything. Proceeding in this way also in the description of numerical relational structures – and this, in accordance with the representation theory of measurement, must be done – we shall thus have to transform the numerical operation of addition, too. Yet, are we then permitted at all to talk about numerical operations? In any event, Suppes does so quite customarily and, in analyzing the systems of extensive measurement, also speaks about the operation of concatenation. Therefore, either he is inconsistent or uses the confusion between a syntactical description and a semantical meaning of the concept of an operation in order to support his initial aim, even if the arguments are faulty. The reason he gives for his defence – namely, that the empirical operation of concatenation, unlike the numerical operation of addition, cannot satisfy the condition of closure, since for the measurement of length one cannot postulate the existence of lengths having arbitrarily large sizes – is insufficient. A similar argument might be given, as has already been remarked, regarding each component of an empirical relational structure which will never precisely satisfy all the properties of the corresponding numerical correlates.

Another specification of the representation theorem refers to the variability of a numerical assignment. Similarly, as there can be given several axiomatizations of one and the same system, the methodologists of extra-physical measurement accept several numerical functions ϕ which satisfy characteristic properties of some empirical relational system. Because a numerical assignment may lead to distinct representations – not only to additive, but also to multiplicative or exponential ones – for a definite empirical structure more than one representation theorem can be proven.[37] This ambiguity refers to both the representation theorems derived for fundamental measurement and those for derived measurement.

According to Suppes and Zinnes, an empirical relational system of extensive measurement $\mathscr{E} = \langle E, Q, S \rangle$ can be mapped either onto the numerical relational system $\mathscr{N} = \langle N, \leqq, + \rangle$, or onto the system $\mathscr{N}^* = \langle N^*, \leqq, \cdot \rangle$. Their justification links this possibility with the transitivity of a homomorphic (isomorphic) mapping:

If we can homomorphically map an empirical system $\mathscr{E}$ onto the numerical system $\mathscr{N}$, and if we can homomorphically map this "additive system" onto the multiplicative system $\mathscr{N}^*$, it follows from this that $\mathscr{E}$ can also be mapped onto $\mathcal{N}^*$. "From a mathematical standpoint, the representation theorem based on $\mathcal{N}^*$ is as valid and useful as the one based on $\mathcal{N}$, so that there is no basis for interpreting the operation $\circ$ as an intrinsically addition operation rather than, say, a multiplication operation."

The possibility of this choice, which is viewed by Suppes and Zinnes as a conventional affair, influences only the pertinent uniqueness theorem: in the first case it will be a matter of a similarity transformation, in the second case – of an exponential transformation.

Although Suppes and Zinnes formulate their conclusions "from a mathematical standpoint", this relativization has no restrictive function within their formal doctrine of measurement. The meaningfulness of a numerical representation of fundamentally measured magnitudes rests primarily on the manner in which an empirical relational system can be directly numerically mapped and on what can be inferred from this numerical mapping as to the characterization of the respective empirical system. Of course, secondarily, every numerical system can be further transformed. Instead of rational numbers we might, for example, choose their logarithms (decadic or natural) too. Since we do not make such transformations for arbitrary measurements, it cannot be inferred from this that there would really exist empirical counterparts of these numerical values and relations between them such that they could be operationally materialized.

The arbitrary nature of a numerical mapping is being justified also by the conventional character of the empirical interpretation of the abstract operation of concatenation. This argument is illustrated by the linear and orthogonal interpretation of length measurement, as it is discussed in the work of Ellis; it also rests on Campbell's considerations as to the parallel and series connections of conductors.[38]

Besides the usual empirical interpretation of the operation of concatenation – joining two measured objects x and y one after another so that $(xSy)\,K\,z$ – with the numerical mapping

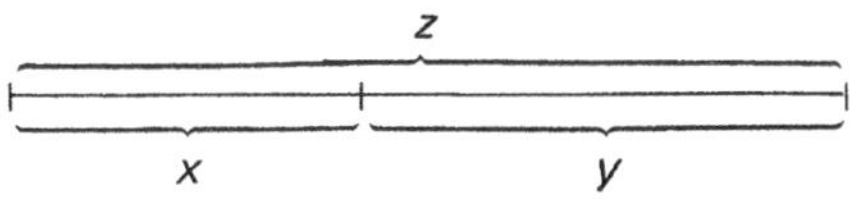

$a + b = c$, Ellis, furthermore, takes into account an orthogonal interpretation based on "right-angled addition". In this interpretation two measured objects x and y are linked so that they form a right angle,

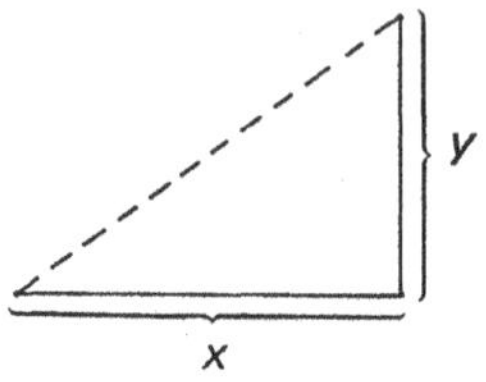

and this leads in the numerical mapping to the expression $a + b = \sqrt{(a^2 + b^2)}$. Concerning this interpretation, it can be geometrically shown that it satisfies all axioms put down for length measurement with a linear interpretation of the operation of concatenation. Therefore, Ellis holds them to be two equivalent variants: the first permits the construction of an additive scale and the second permits multiplicative scales. However, there exists a simple mutual relationship between these scales. After introducing for the orthogonal interpretation one diagonal meter (dim) as a measurement unit of length, we might, on the basis of equations

$$n \text{ dim} = \sqrt{(n)} \text{ m} \quad \text{or} \quad n\text{m} = n \text{ dim}^2$$

transform the scale values of one scale into the scale values of the other.

What is the relationship between these different interpretations of the operation of concatenation? If we take into account the above-mentioned transformation relations and also consider the ordinary practice of an indirect fundamental measurement of distances, in which we likewise use our knowledge of geometrical laws, we cannot agree with the view that we are dealing here with "totally different interpretations of concatenation".[39] Even if it had been the case, they could not have referred to a fundamental measurement of one and the same magnitude. Since this conclusion has already been embraced by Campbell himself, the theorists of extraphysical measurement refer to him in this connection without justification.

To Campbell, it is quite evident that different empirical interpretations can be given to numerical addition within the framework of a definite system, for example, as parallel and series connections of conductors. However, these interpretations can never refer to one and the same magnitude. And, indeed, through the series connection of conduction we measure resistance, whereas by means of the parallel connection we measure conductivity. In this case too, numerical values of resistance may be transformed into numerical values of conductivity and vice versa, for one magnitude is inversely proportional to the other.

The assertion that more than one representation theorem may be demonstrated for one and the same magnitude leads to even more controversial consequences in considering derivatively measured magnitudes. From the standpoint of this position, we could indeed regard, for instance, the additive representation *mass* + *volume* and the multiplicative representation *mass* · *volume*, as two equivalent representations, regardless of the fact that only in the latter case one can provide objective reasons for this.

7. THEORY OF SCALES

A theory of scales can be understood either as a specific variant of the theory of extraphysical measurement, oriented especially to the problems of scaling and scaling procedures, or as a relatively independent component of a general theory of measurement. The second alternative is to be preferred to the first one; this follows from our analysis of the concept of a scale with respect to the concepts of measurement and magnitude, as well as from our exposition of the theory of measurement. Yet even in this case we cannot avoid mentioning some problems which have been outlined in connection with the acceptance of the first alternative. Obviously, we shall not repeat what has already been said in the same or only a slightly modified form; nevertheless, it will not be without interest for the confrontation of different doctrines to point out some mutual connections.

To complement the foregoing analyses of the theoretical aspects of measurement, it will be sufficient to concentrate our attention in this chapter on two thematic areas: (i) the classification of scale types, which constitutes the most elaborated part of the theory of scales, and (ii) the problems of the invariance of a scale form with regard to various transformations, which are variously solved within the formal theory of measurement by the derivation of uniqueness theorems.

7.1. THE CLASSIFICATION OF SCALE TYPES

All familiar classifications of scale types – or classifications of scales, as it is often inexactly expressed under the influence of overlooking the difference between individual scales and classes of scales the elements of which are scales of the same form – are contingent upon the ways of conceiving the extension of the concept of measurement. Wider and narrower classificational systems might then be discerned accordingly. In the wider classifications, developed particularly within the framework of extraphysical measurement,[1] all scales – non-metric and metric – are held to be scales of measurement and among non-metric scales are usually also included those scales which are obtained by a mere numeral assignment. In the narrower classificational systems[2] there is concern about the difference between numbering, scaling, and measuring, and only those non-metric scales which satisfy the conditions of topologization are admitted as being non-metric scales.

The *wider classifications* have their origins in Stevens' system which has been consciously designed so that it would satisfy certain classificatory criteria. Stevens strives to build up his system of scale types on the basis of the criteria of *empirical operations, mathematical structure of scales*, and an *applicable statistics* with regard to the elaboration of obtained numerical data. Considering these three criteria, only the first two are relevant to the assessment of his doctrine. The use of suitable statistics is already an affair of a practical nature and it is not immediately connected with the construction of particular scale types.

On the basis of these criteria, or rather with regard to his intuitive ideas, Stevens distinguishes four types of scales: nominal (name), ordinal (rank-order), interval, and ratio. In his classificational system, the scale types are arranged hierarchically, in an ascending order, beginning with the nominal scale, which is the weakest, and ending with the ratio scale, which is the strongest. Every next scale presupposes the previous ones.

The *nominal scale*, as the most elementary scale type, is based on any assignment of numerals in the function of mere naming. The nominal assignment of numerals serves the numbering of individual objects for the sake of their identification, or the numbering of classes, where each member of a class is assigned the same numeral. In either case according to the rule, "do not assign the same numeral to different classes or different numerals to the same class."[3]

The basic empirical operation which defines this scale type is "determination of equality", so that only the first correspondence rule of topologization (formula (6.3–15)) is to be considered for the assignment of scale values to empirical objects. Hence, by means of scale values of a nominal scale, one might decide only the question of whether two objects belong (or do not belong) to the same class.

The mathematical structure of this scale type is defined by the permutation group

$$x' = f(x),$$

where x and x' denote, in this case, arbitrary numerals and $f(x)$ expresses a mutual one-to-one substitution.

This scale type remains invariant under every change of a numeral assignment which preserves the identity of a class, i.e., under the general substitution or permutation group, the so-called symmetric group of transformations.

Since the form of nominal scales is not subjected to a change if we apply to them some mutually one-to-one transformation, this scale type is sometimes designated by the term *absolute scale*.

The *ordinal scale*, which presupposes a natural ordering of objects with respect to some property, is defined by two empirical operations: "determination of equality" and "determination of the relation 'greater or less'". These operations permit the application of both correspondence rules of topologization, (formulae (6.3–15) and (6.3–16)).

The mathematical structure of ordinal scales is characterized by the isotonic (i.e., order preserving) group

$$x' = f(x),$$

where $f(x)$ is again, in this case, any increasing or decreasing monotonic function. Therefore, the form of ordinal scales remains invariant under every monotonic transformation.

The *interval scale* differs from the previous scale types in that it requires a stipulation of a measurement unit and of an arbitrary origin, and that all arithmetical operations are admissible for it, as far as they refer to the difference between the pairs of scale values, although not all of them are to be operationally significant.

Nevertheless, the basic empirical operation, specific for this scale type, does not correspond to this general characteristic. "Determination of equality of intervals or differences" permits only such a numerical assignment in which the equality of differences between the pairs of scale values expresses the equality of intervals between the pairs of empirical objects with respect to some property, hence the same as the correspondence rule of topologization, (formula (6.3–15′)). However, this is not yet sufficient grounds for regarding this scale type as a "quantitative" one, in the customary sense of the term.[4]

With regard to a mathematical group structure, the interval scale is characterized by the general linear group

$$x' = \alpha x + \beta \quad (\alpha > 0), \tag{7.1–1}$$

where α denotes a measurement unit and β a scale origin. The interval scale is thus invariant under every positive linear transformation.

The *ratio scale*, from a formal standpoint, is indeed an interval scale with a natural origin, i.e., with an additive constant $\beta = 0$. Its characteristic empirical operation, supplementing the empirical operations assumed for the previous scale types, is "determination of equality of ratios". All arithmetical operations may be applied to the scale values of a ratio scale, though not all of these operations are to be operationally significant, as is the case of the operation of addition.[5] The numerical assignment, by means of a ratio scale, is invariant under any change of the basic measurement unit. On the other hand, the change of the origin or its arbitrary stipulation leads to a change in the interpretation of ratios; this scale type is then transferred into an interval scale.

Stevens' ideas of the empirical operation of *determination of equality of ratios* refer to scale values rather than to the elements of the empirical set E. If we assign to empirical objects x, y scale values na and nb, where n is a basic measurement unit, the ratio of the scale values na/nb, assigned to the *empirical ratio* x/y, will be equal to any other ratio of the scale values $n'a/n'b$, where n' is some multiple or portion of the measurement unit n; then it will be possible to assign this new numerical ratio to the *empirical ratio* x/y.

The mathematical group structure of ratio scales is expressed by the similarity group

$$x' = \alpha x \quad (\alpha > 0), \tag{7.1–2}$$

where α denotes a measurement unit. The form of this scale type thus remains invariant with regard to every similarity transformation.

The hierarchy of scale types in Stevens' doctrine, containing moreoever the so-called *logarithmic interval scale*, can be collectively summarized in the following schema:[6]

nominal scale

$x' = f(x)$

(permutation group)

ordinal scale

$x' = f(x)$

(isotonic group)

linear interval scale

$x' = \alpha x + \beta$

(linear group)

logarithmic interval scale

$x' = kx^n$

(exponential group)

ratio scale

$x' = \alpha x$

(similarity group).

If we are to concede that the "determination of equality", which refers to individual elements, their intervals, and ratios, is indeed an empirical operation (as Stevens believes), we would have to show that the notions of equality, interval, and ratio are primarily empirical concepts. One would also have to explain clearly what it is on which the operative character of these operations rests, and how these operations are correlated with the rules of a numeral or numerical assignment. Yet, in reference to these principal problems, we do not find in Stevens any satisfactory explanation. Not only do we miss concrete specifications of these concepts, but also a general definition of their properties. With the exception of nominal scales, the rules of numeral assignment are not explicitly formulated either. Generally conceived operations of the "determination of equality" are, in fact, only abstract correlates of the numerical characteristics of scales and cannot serve as the empirical point of departure for the process of measurement.

Stevens' classification of scales – as is evident from the above-mentioned schema – is therefore not supported by a criterion of empirical operations, but merely by the criterion of the mathematical structure of scale types – by their group structure, conditioning admissible transformations under which the form of the scale does not change. At any rate, this conclusion has been reached, even by the followers of Stevens and, in particular, by the adherents of the formal theory of measurement.

The first reinterpretation of Stevens' theory is to be found in the work of Coombs. On his view, scale types are mathematical models representing hierarchically ordered levels of measurement, "capable with an increasing force to transmit information concerning empirical data". The foundation of his classification consists of the criterion of the analytic applicability of arithmetical formulas which unifies the mathematical component of Stevens' criterion of empirical operations with the criterion of mathematical structure. Then, on the ground of this criterion, individual scale types are characterized only by arithmetical formulas, for example, interval scales by the formulas

$$a \gtreqless b \quad \text{and} \quad |a-b| = |c-d|,$$

of which it can be proven[7] that there correspond to them analogical formulas for transformation functions which fulfill Stevens' requirement of the invariance of a scale form.

Since Coombs quite explicitly emphasizes only the mathematical aspect of the theory of scale types, he is not at all interested in the operational realizability of particular operations of the "determination of equality". Of course, by employing his classification of scale types, Coombs pursues different methodological goals. By means of introducing further scale types of a lower level, he orients his doctrine rather towards the needs of scaling, and not towards the purposes of a variant formulation of the theory of measurement in Steven's sense.

Coombs first of all augments the classification of scale types by so-called *partially ordered scales*, which are distinguished from so-called *simply ordered scales* (ordinal scales in Stevens' view) by the fact that the relation *more* (or *less*) holds only for some ordered elements. A much more essential modification is the introduction of so-called *composite scales*, which are classified with regard to the classes of objects themselves and with regard to the distances between them. Since Coombs discerns three basic types of non-metric scales (nominal, partially ordered, and ordered), he differentiates between nine types of composite scales altogether which are uniformly named so that the first name refers to objects and the second to the distances between them, for instance, the so-called nominal-nominal, nominal-partially ordered, and nominal-ordered scales. Of course, if he considers only objects themselves, these composite scales are identical with nominal scales.

Like to Coombs' classification, Torgerson's classification rests on the assumption that scale types are to be regarded as formal, mathematical models which must satisfy at least one of the general characteristics of numbers (Section 4.3). The first property – ordering – must be satisfied by every scale type; the other two properties – distance and origin – then define the sufficient conditions for the characterization of the specific properties of particular scale types. Torgerson's classification then contains these scale types: ordinal (without distance and origin), ordinal with a natural origin, interval (without distance and without a natural origin), and interval with a natural origin.

In this classification we do not find a nominal scale any more. The ordinal scale with a natural origin is only a variant of an ordinal scale and the interval scale with a natural origin is identical with the ratio scale.

A consistent reinterpretation of Steven's classification from the standpoint of the formal theory of measurement, has been offered only by Suppes and Zinnes. According to their view, the concept of a scale type is to be defined exclusively by means of classes of numerical assignment. Therefore, only the admissible transformation properties of a numerical assignment, and not, however, empirical operations, are indeed relevant to the differentiation of various scale types. Allegedly, empirical operations have no influence either on the determination or on the construction of scale types, the number of which is theoretically unrestricted, just as the number of possible formal transformations is unlimited.

However, even within this theory of measurement several basic kinds, which are also assumed to be empirically significant, are usually distinguished. Of course, they need not be of the same scale types. Thus, for example, besides nominal, ordinal, interval, and ratio scales, characterized by analogical invariant transformations as in Stevens, Pfanzagl also mentions the difference scale which is unambiguous up to "shifts"

$$x' = x + \beta, \tag{7.1–3}$$

i.e., up to some additive constant β. On the other hand, with respect to various numerical structures of the kinds of measurement, Kanger distinguishes (i) ordinal, (ii) halving, (iii) interval, (iv) difference, and (v) ratio (i.e., proportional) scales. These scale types are characterized by the following numerical expressions:

(i) $a \leqq b$
(ii) $b + b \leqq a < c$
(iii) $|a - b| \leqq |c - d|$
(iv) $a + b \leqq c + d$
(v) $a \cdot b \leqq c \cdot d.$

Besides these scales, he admits, furthermore, other possible types, for example, so-called semi-ordinal scales.

In the *narrower* classificational systems, scale types are classified according to the differentiation of the topological and metric conditions of measurement, or in correlation with extensive and intensive magnitudes, or, finally, in connection with the differentiation between scaling and measuring.

Insofar as the respective magnitudes satisfy only the conditions of topologization, and their scale values are merely ordinal numbers, *non-metric* or *topological* scales (i.e., the ordinal scales of Stevens' system) correspond to them. If they satisfy the conditions of metrization, too (i.e., if their scale values are cardinal numbers), *metric* (or additive) scales correspond to them.

The elimination of nominal scales from the class of non-metric scales will also be acceptable to many representatives of the wider doctrine of measurement. Could not the class of metric scales be restricted in a similar manner? Do we have to take for independent scale types, besides the ratio (or proportional) scales, also the interval scales, or perhaps the difference and halving scales as well? In discussing this problem it will suffice to focus our attention only on the interval scales.

The relevance of the discrimination between interval and ratio scales, insofar as it is justified by the distinction between an arbitrary and a natural scale origin and put into connection with the distance function, has already been deemed rather doubtful in view of the results which are arrived at in elucidating the notions of the scale origin (Section 4.2) and distance (Section 4.3). Therefore, there is no need to repeat them.

Hence let us investigate another argument, which rests on the criterion of the mathematical structure of scale types and is concretely illustrated through the customary example of temperature scales. When comparing the mutual relationship between the temperature scales of Celsius and Fahrenheit, it is sufficient to use the commonly familiar transformation formula

$$°\mathrm{F}(x') = 9/5\,°\mathrm{C}(x) + 32, \qquad (7.1\text{–}4)$$

where $°\mathrm{F}(x')$ and $°\mathrm{C}(x)$ denote scale values, or the formula

$$°\mathrm{C}(x') = 5/9\,(°\mathrm{F}(x) - 32). \qquad (7.1\text{–}5)$$

Since Fahrenheit's scale can be transformed into the scale of Celsius and vice versa, their mutual relation can be represented quite generally by a linear transformation

$$x' = \alpha x + \beta \quad (\alpha > 0). \qquad (7.1\text{–}1)$$

The possibility of the linear transformation of one of them into the other is regarded as a sufficient condition for interpreting these scales as interval scales.

This argumentation is only apparently demonstrative. In analogically comparing the scales of Celsius and Réaumur we arrive at a different result. These two scales are transformable on the basis of the formulas

$$°\mathrm{C}(x') = 5/4\,°\mathrm{R}(x) \qquad (7.1\text{–}6)$$

and

$$°\mathrm{R}(x') = 4/5\,°\mathrm{C}(x), \qquad (7.1\text{–}7)$$

so that their mutual relationship can be expressed by means of the similarity transformation

$$x' = \alpha x \quad (\alpha > 0). \qquad (7.1\text{–}2)$$

If we take group properties into account, then these scales must be interpreted as ratio scales.

The conflict between these interpretations is obvious: in the former case, in comparing the scales of Celsius and Fahrenheit, we had to deal with two distinct points of reference – two scale origins, whereas in the latter case, in comparing Celsius' and Réaumur's scales, we had to operate only with one such point.

The comparison of two different temperature scales has thus led to incompatible results. If the form of one and the same scale, namely Celsius', is in one case invariant as to the linear transformation, then it is to be interpreted, in the sense of the mathematical criterion of Stevens' classification, as an interval scale. Yet, if in another case it is invariant as to the similarity transformation, it is to be interpreted as a ratio scale.

The difference between the distinct invariant transformations which characterize both these scale types depends from a purely mathematical standpoint on whether $\beta = 0$ or $\beta \neq 0$ is the case. Because either of these scales can be generally characterized by means of the same linear transformation, we could, in turn, view the interval scale as the basic one and the ratio scale only as its limiting case, i.e., as an interval scale with the additive constant $\beta = 0$.

If the transformation properties are not sufficient for a discrimination between interval and ratio scales, we might again consider the differentiation between an arbitrary and a natural scale zero. In connection with the transformation criterion, this distinction depends rather upon our procedures of comparison – whether we compare scales with the same or different origins, disregarding at the same time the fact of their arbitrary or non-arbitrary character. Although Réaumur's and Celsius' scales have an arbitrary origin, they are nevertheless mutually transformable by similarity transformations.

The objection may be raised against this interpretation that it distorts Stevens' argument concerning the possibility of measuring temperature, by using two different scale types with distinct transformation properties, contingent upon the choice of an arbitrary or a natural scale zero.

Not only will the discussion of this argument show the limitation of the transformation criterion, but it will also elucidate one of the probable reasons why Stevens introduced the terms 'interval' and 'ratio' scale.

Let us choose an ideal gas as a uniform thermometric material. This is possible, of course, only theoretically, for different temperature scales are always graduated according to some concrete thermometric material. However, we take such a course in order to be able to find a uniform measuring device for the comparison of these two different scale types with distinct transformation properties. On the basis of the functional relationship between temperature and volume (under a constant pressure), temperature measured in Centigrades as well as in Kelvin grades, can now be expressed by these equations:

$$t = \frac{V - V_0}{V_{100} - V_0} \cdot 100^\circ \mathrm{C},$$

$$T = \frac{V}{V_0} \cdot 273.16\,\mathrm{K},[8]$$

where V denotes the given volume, V_{100} the volume at the temperature 100° C, and V_0 the volume at the temperature 0° C; t denotes the temperature in °C, and T the temperature in K. In the first case, we are dealing on one side of the equation with *intervals* or *differences* between volumes, while in the second case, with their *ratio*. In either case, it suffices, therefore, to investigate volume – at one time, the difference between volumes, and at another, their ratio. Hence, at one time we can talk – *per analogiam* – about an 'interval' scale, and at another, about a 'ratio' scale.

Can we, however, draw from this relevant consequences concerning the character of the measured magnitude? Is the analogy between the ratio of volumes and ratio scales, on the one hand, and between the intervals or differences between volumes and interval scales, on the other, sufficient for the discrimination of two different scale types with distinct transformation properties? So far we have considered only the consequences which follow from this differentiation for the assessment of the relationship between the transformation properties of scale types and the properties of the measured magnitude, namely of temperature. Since different collections of basic arithmetical magnitudes have a significant empirical interpretation for the interval and ratio scales, we would have to assume, in moving from one scale type to another – as we have, at any rate, previously mentioned – a change of qualitatively determined, objectively existing properties of temperature: on the interval scale temperature it would be a non-extensive magnitude, while on the ratio scale it would be an extensive magnitude. Of course, this is an evident logical inconsistency.

If we measure temperature in K, as we are committed to by the legislated measurement units, we shall have to view it, in accordance with Stevens' doctrine, as a magnitude measurable by means of a ratio scale. If the temperature is measured in °C, we must then interpret it as a magnitude measurable by means of intervals.

However, what other illustrations of interval scales which satisfy the requirement of a measurement unit can we provide in addition? It is true that in the theory of extraphysical measurement, where a great significance is assigned to this scale type, various cases of an interval measurement are given, for example, the measurement of expected utility, yet without fulfilling the required condition of a measurement unit. Nonetheless, if this condition is not met, an interval scale could hardly be regarded as a metric scale.

If we refuse to accept interval scales as independent scale types, it will not be expedient to speak of ratio scales either. Consequently, the system of classical scale types, as elaborated by Stevens, will be reduced to two types, namely to *non-metric* and *metric*, which at the same time represent two distinct, mutually connected empirico-mathematical methods of *scaling* and *measuring*.

This doctrine can be supported by still another weighty argument, which follows from the previous exposition of topologization and metrization. The correspondence rules of topologization express the same properties of weak ordering, whether they refer to individual elements or pairs of elements from the set E. The correspondence rules of metrization, which are primarily conditioned by the existence of a measurement unit and only secondarily by the empirical realizability of the condition of additivity, can analogically also be formulated for binary relations of the type K and P, or Q, or for quaternary relations of the type I or $\boldsymbol{D}$. Since neither the 'distance function' nor the 'equal spacing' are sufficient substitutes for a measurement unit, from this standpoint we might also consider interval and difference scales as only specific variants of ordinal scales, i.e., scales of a non-metric type. Consequently, the real reason for introducing these scales into the theory of scales vanishes altogether.

7.2. SCALE TRANSFORMATIONS AND THE UNIQUENESS THEOREM

On the basis of the foregoing exposition, we can now assess the appropriateness of the concept of transformation invariance for the differentiation of various scales. In this respect we shall be especially interested in the issue of whether scale transformations refer to scale types or to individual scales, as well as whether one can draw from the transformation properties relevant conclusions concerning the nature of the numerical mapping, i.e., concerning the properties of empirical objects as well.

Considering the discussion of the mutual relationship between temperature scales, we must concede that transformation properties do not refer to individual scales but only to the formal properties of classes of scales – to the scale types. Their significance rests on the fact that they help to decide – by comparing mutual formal properties of individual scales – whether these scales belong or do not belong to one and the same scale type. These properties are properties of classes of scales, and not those of individual scales. They are relevant only for grasping mutual relationships between scales within one scale type, and serve merely the transformation of one element into another one from the same class of scales. This is also the reason why these properties can be applied in different classificational systems – not only in the wider ones but in the narrower too – which are based on differentiated classificatory criteria.

Nor do the transformation properties display a necessary relationship to the species and character of the measured magnitudes. A change of the formal properties of scale types cannot lead to a change of qualitatively determined magnitudes. Should we assume the contrary, we would have to admit the absurd consequence that temperature, which is judged to be measurable both on the interval scale and the ratio scale, is at the same time a metric and non-metric magnitude.

The mathematical properties of scales are therefore not sufficient for defining the measured magnitudes, just as this cannot be achieved merely by considering the operational characteristics of the measuring procedures. In either case one must take into account both the methodological aspects and the purely practical, applicational points of view. Nor should we omit the unifying function of scientific theories, conditioned by the historical level of our knowledge.

Since the transformation criterion is relevant only for the scale types, Stevens' classification is a classification of those. For this reason it cannot be simultaneously based on some empirical criterion which would be relevant only to the differentiation of individual scales. At any rate, Stevens always talks only very abstractly about "determination of equality", without specifying this determination in any concrete way. As soon as we realize that a scale type is a class of scales having the same form, this conclusion follows immediately from the familiar difference between a class and its elements.

From the differentiation of *scales* and *scale types* there also follows a similar consequence for all other classificational systems of scales, which again merely refer to classes of scales. Because the scale types are not – unlike the individual scales – the results of scaling or measuring, but equivalence classes that are formally defined by means of invariant transformations, neither empirical relations and operations nor their operational realizability can be relevant to them. Of course, if we are to assess the adequacy of a numerical mapping, the ascertainment of the formal properties of concrete empirical relations and operations, as well as their categorization, are necessary indeed. Only in this mediated way are the categorizations of the formal properties of empirical relations and operations also tied up with the differentiation of various scale types.

The transformation invariance of a scale form may also be alternatively formulated with the help of uniqueness theorems, which are being derived – analogically to the theorems of representation – from axiomatically defined relational structures.

The requirement of uniqueness of a numerical mapping is already contained in the assumption of non-ambiguity of measurement in the classical theory of measurement. Namely, Campbell regards every measurement, insofar as it meets the condition of additivity, as unique "except for the arbitrary choice of the unit."[9] This means that the numerical values assigned to the quantity of the measured magnitudes depend on the choice of the measurement unit and change according to each alteration of this unit.

To every scale value a expressed in the measurement units j, one may assign exactly one scale value a' expressed in the measurement units j'. This mutual relationship can be generally characterized by means of a similarity transformation (formula (7.1–2)), where the transformation coefficient α always denotes a measurement unit.

The proper conception of uniqueness "up to some transformation" and the manner of its proof in the formal theory of measurement are again tied up with the von Neumann–Morgenstern axiomatization of expected utility.[10] In this case, the linkage is to the second basic theorem, to the *uniqueness theorem* (A : W):

"For any two mappings $\mathrm{v}(w)$ and $\mathrm{v}'(w)$ possessing the properties (i) and (ii) – in the sense of the existence theorem (A : V) – we have

$$\mathrm{v}'(w) = w_0\ \mathrm{v}(w) + w_1,$$

with two suitable but fixed [numbers] w_0, w_1 and $w_0 > 0$," which expresses the idea that "utility is a number up to some linear transformation."

To every scale value assigned to utility, one might therefore find some other scale value on the basis of linear transformations ((formula 7.1–1)), which, in the formal theory of measurement, also comes to be expressed in the form

$$\phi'(x) = \alpha\phi(x) + \beta, \qquad (7.1\text{–}1')$$

where ϕ' and ϕ are two real-valued functions.

However, as it is explicitly expressed in the von Neumann–Morgenstern formulation, it is not requested here that the transformation coefficient w_0 should denote a measurement unit and w_1 a scale origin. Nevertheless, as the point of departure, one finds here an analogy with the interval measurement of temperature. However, if this interpretation is lacking, the transformation formula in question does not have the same meaning as the condition of invariance for the temperature scales of an interval scale type.

Let us therefore consider the question of whether the justification of measurability of the expected utility could really rest on the analogy referring to the temperature measurement – an analogy which is also commonly used in respect to other non-extensive magnitudes. In discussing this problem, it will suffice to investigate in the measurements of utility how those basic properties of interval scales are realized that are satisfied by all temperature scales of the interval type, i.e., *the constant unit of measurement, an arbitrary origin of the scale*, and *the invariance of a scale form as to every linear transformation*.

We shall begin with the first property, which is the most essential condition of measurability. Whereas for temperature, we have at our disposal a conventionally established, yet objectively reproducible, constant unit of measurement, for utility there has not been found any acceptable pseudo-unit. It is true, that some writers[11] solve this situation simply by taking the so-called 'util' for the measurement unit of utility without explaining at the same time how they define it or how they imagine its empirical reproducibility operates. If such justification is lacking, the introduction of a measurement unit is only fictitious and, at any rate, cannot contribute to arriving at the postulated analogy with measuring temperature.

Some writers define 1 util as a utility of 1 dollar. Of course, this evidently conflicts with the theory of expected utility.[12] Others choose instead a more neutral attitude. They express the measurement unit of utility, just like the origin of a 'utility scale', by any real numbers, shunning any further designation.

Let us concede that we can indeed choose any pairs of numbers, for example, 13, 105; 3, – 39. What are the reasons for this, if we do not regard one of them as the measurement unit and the other as the origin of a scale? Obviously, without this interpretation, one cannot justify the linear nature of the scales of an interval type, for by means of a pair of numbers α, β $(\alpha > 0)$ every scale value x may be unambiguously transformed into a scale value x', perhaps even on a nominal scale. The choice of two numbers α, β in the formula of linear transformation is thus in itself insufficient to warrant our conclusion concerning the measurability of some magnitude having scale values of interval scales.

At any rate, this arbitrariness is restricted in the operationalization of a utility function by the fact that the preference ordering of utilities is reduced to a certain interval which is bounded by the least and by the most desirable alternatives x and y, to which numbers a, b are being assigned so that $a < b$, usually numbers 0 and 1. Can these numbers be regarded as a measurement unit and a scale zero? Concerning this problem we find, even in one and the same writer, two incompatible answers:[13] at one time he takes for a measurement unit the number 1, yet at another time he takes the difference between the numbers 0 and 1. With respect to the assumed analogy with measuring temperature, none of these explications is acceptable.

Even the arbitrariness of the origin of a 'utility' scale should not be interpreted in the same sense as the arbitrariness of the scale zero of temperature scales. We may select any number a to serve as the origin of a 'utility' scale, as long as it satisfies the inequality $a < b$; this choice is theoretically unlimited. In the case of temperature scales, the choice of the numerical value of the origin is very narrowly restricted and depends upon an entire range of context-relevant circumstances. Among other things, it depends upon the properties of the thermometric material. The origin of 'utility scales' is furthermore defined so that it numerically represents the least desirable element of a given preference ordering. On the other hand, the origin of temperature scales of an interval type does not correspond to the least value of temperature. Hence, the origin of 'utility scales' at one time has a more arbitrary nature, and yet at another, it is in turn more "natural" and even distinguishes itself by a similar property, such as the absolute origin of Kelvin's scale. In a standard numerical mapping, however, the preference is given rather to the so-called canonical form, in which 0 denotes the beginning and 1 the end of a utility interval.

It remains, therefore, to assess in what measure a third property is being satisfied, from which it is inferred that the expected utility is a magnitude measurable at least on an interval scale. According to our view, as concerns 'utility scales', one cannot meaningfully interpret the transformation formula which keeps the scale form of interval scales invariant. The criterion of an invariant transformability is a characteristic mark of interval scales exactly only in the case if we can assign to the transformation constants a semantically justifiable and operationally realizable meaning: α must represent the measurement unit and β the origin of a scale. However, as it is already evident from the previous exposition, for two distinct 'utility scales' such an interpretation is not to be considered at all, especially for the coefficient α.

Without this interpretation, the linearity of a utility function rests only on a superficial analogy with the linearity of temperature, from which, however, it cannot be inferred that the expected utility would be a measurable magnitude. Insofar as no measurement unit can be established for this magnitude, it cannot be held to be measurable in the strictest sense of the term. Neither can the measurability of expected utility be justified by the fact that the numerical values of a utility function can be linearly transformed into other numerical values. And this is exactly the only sense through which one can understand the notion of linear transformability for the expected utility.

In the von Neumann–Morgenstern doctrine, this interpretation rests on the numerical transformation formula

$$\rho' = \phi(\rho) = w_0 \rho + w_1, \tag{7.2–1}$$

where $\rho = \mathrm{v}(u)$ and $\rho' = \mathrm{v}'(u)$ are two numerical mappings and w_0 and w_1 ($w_0 > 0$) are two arbitrary fixed numbers. If we apply them to the numerical components of the basic formulas which express both properties of an expected utility, we obtain the following expressions:

$$\mathrm{v}(u) > \mathrm{v}(v) \quad \text{implies} \quad \mathrm{v}'(u) > \mathrm{v}'(v) \tag{7.2–2}$$
$$\mathrm{v}'(\gamma\,\mathrm{v}(u) + (1-\gamma)\,\mathrm{v}(v)) = \gamma\,\mathrm{v}'(u) + (1-\gamma)\,\mathrm{v}'(v). \tag{7.2–3}$$

The role of the transformation formula (7.2–1) may be elucidated by the following numerical model. Let $\mathrm{v}(u) = 4$, $\mathrm{v}(v) = 2$, $\mathrm{v}'(u) = 18$, $\mathrm{v}'(v) = 10$, and $\gamma = 0.5$. Then clearly it holds that

$$4 > 2 \quad \text{implies} \quad 18 > 10,$$

yet, contrary to the formula (7.2–3), we obtain now an inequality

$$4 \cdot 0.5 + 2 \cdot 0.5 \neq 18 \cdot 0.5 + 10 \cdot 0.5,$$

or

$$3 \neq 14.$$

In order to attain the required equality, it is sufficient to apply the following modification, using the formula (7.2–1)

$$(4 \cdot 0.5 + 2 \cdot 0.5) \cdot w_0 + w_1 = 18 \cdot 0.5 + 10 \cdot 0.5,$$

or

$$3w_0 + w_1 = 14,$$

which facilitates the calculation of numerical values for the constants w_0 and w_1:

$$\begin{array}{ll} 4w_0 + w_1 = 18 & \\ 3w_0 + w_1 = 14 & 12 + w_1 = 14 \\ \hline w_0 = 4 & w_1 = 2 \ . \end{array}$$

Let likewise $\mathrm{v}(u) = 10$, $\mathrm{v}(v) = 5$, $\mathrm{v}'(u) = 100$, $\mathrm{v}'(v) = 30$, and $\gamma = 0.8$. Then it obviously holds that

$$10 > 5 \quad \text{implies} \quad 100 > 30$$
$$(10 \cdot 0.8 + 5 \cdot 0.2)\, w_0 + w_1 = 100 \cdot 0.8 + 30 \cdot 0.2$$

where, according to the previous manner of calculation, $w_0 = 14$ and $w_1 = -40$.

It is evident from these transformations that the numerical constants w_0 and w_1 depend upon the numerical values $\mathrm{v}(u)$, $\mathrm{v}'(u)$, $\mathrm{v}(v)$, $\mathrm{v}'(v)$, and γ. Yet, how are these numerical values obtained? The transformation of some numerical expression into another one is an entirely secondary matter. What indeed would be needed is to find some objective way to establish initial numerical values which are assigned to the utilities of the investigated alternatives. In our search for the solution of the primary problem of measurement, the numerical constants w_0 and w_1 are of no avail. They really are arbitrary numbers that cannot be interpreted as a measurement unit and an origin of a scale. Nevertheless, in transforming one temperature scale into another, the meaning of these transformation constants α and β is exactly determined; they are given antecedently and they also immediately condition the size of numerical values which are assigned to different degrees of temperature.

However, in the formulations of the uniqueness theorem, this essential difference tends to be forgotten – although such an omission is not rightful, yet the reasons for it are quite obvious. From the standpoint of the formal theory of measurement, it is irrelevant whether transformation constants have any meaning at all – whether they are empirically significant. It is only important whether this theorem can be derived from axiomatically formulated definitions which delimit the properties of abstract relational structures, or from other suitable definitions. And it is exactly this second alternative that is the usual point of departure for these seemingly correct proofs, which, however, are in fact suffering from a methodological fallacy of *petitio principii*. Namely, they assume in advance what is only to be proven.

In order to demonstrate that some real-valued function ϕ is unique up to some linear transformation, in the derivation of the uniqueness theorem[14] for the structure $\langle E, D\rangle$ of the interval measurement there are introduced for every x from the set E two functions h_1 and h_2 such that h_1 is a linear transformation of the function ϕ_1 and h_2 is a linear transformation of the function ϕ_2 and also h_1 is identical with h_2. It is evident from these definitions – which, however, are not formulated by means of the primitive notions of the given axiomatic system – that the functions h_1 and h_2 had been introduced so that they are invariant with regard to every linear transformation. Then it really can be easily shown that the function ϕ_1 is a linear transformation of the function ϕ_2, and thus that there exist numbers α and β ($\alpha > 0$) such that for every x in E the following holds

$$\phi_1(x) = \alpha\phi_2(x) + \beta. \qquad (7.1\text{–}1'')$$

It is also evident from this procedure that linear transformations refer to a previously given numerical expression.

On the ground of supplementary, suitably chosen definitions, uniqueness theorems are also analogically demonstrated for further structures of non-extensive measurement. Thus, for example, for measurement by halving (bisection measurement), which is formally defined by another set of primitive notions and axioms, the same uniqueness theorem is being derived as the one for the interval measurement.

The congruence between uniqueness theorems for interval measurement and measurement by halving may be explained by the fact that these theorems – in spite of being derived from different axiomatic systems that are not mutually equivalent – are based on the same formal transformations.

Measurement by halving (measurement by means of a bisection) can be explicated by Ramsey's analysis of the logic of 'partial beliefs',[15] which is the first systematic attempt to constitute the logic of value judgments. The fundamental notions of Ramsey's system are the concepts of preference ordering, numerical mapping of desirability onto the probability interval [0, 1], and the concept of degree of belief in different statements, in particular in the so-called neutral statement with the numerical value $p = 1/2$. By means of an ethically neutral statement and the concept of indifference as to two choices, one can accomplish by a successive bisection of the interval of desirability an equal spacing of this interval.

For the purpose of our analysis, it will suffice to mention only equivalent transformations of preference ordering which can be expressed, based on R. C. Jeffrey's interpretation, by a certain matrix of desirability of the type 2 × 2. A matrix of desirability M_D characterizes the desirability of the results of definite activities realizable under certain circumstances by means of the qualitative attributes 'very desirable', 'less desirable' . . . 'entirely undesirable', which are designated either by numbers from the interval [0, 1] or by the integers . . . –2, –1, 0, 1, 2, . . . , where negative numbers are to express "evils", positive numbers "goods", and zero the neutral value. Although the assigned numbers merely express ordering, one deals with them as if they were cardinal numbers.

If two different matrices lead to the same preference ordering, they are held to be equivalent. The ascertainment of this equivalence is founded on their mutual transformability, either with the help of a *general* or by means of a special *linear* transformation.

In the first case, every element of one matrix will be multiplied by the same positive number or it will be augmented (reduced) by the same arbitrary number; generally expressed

$$d \in M_D \Rightarrow d^* \in M_D, \tag{7.2–4}$$

where $d^* = ad + b,\ a > 0$; hence similarly as in (7.1–1). By means of this transformation, every matrix M_D can be brought to its canonic form M_D^* in which the most desirable element will be designated by the number 1, the least desirable element by the number 0, and all others by the numbers in the interval (0, 1) calculated by means of the respective values of the transformation constants a and b.

Let us assume that somebody is deciding between two activities, "to wash the car" – "not to wash the car", under the conditions, "it will rain tomorrow" – "it will not rain tomorrow", and that he expresses his impressions referring to the weather, his desire to have a clean car, his attitude to the effort connected with this, etc., by the following matrix of desirability

$$M_D : \begin{bmatrix} -100 & 0 \\ -50 & 50 \end{bmatrix}.$$

According to the established procedure, we shall thus assign to the extreme values –100 and 50 the canonical values 0 and 1, and use the transformation expression $ad + b$. In accordance with the canonical values, we can express the least desirable and the most desirable element as follows:

$$-100a + b = 0 \qquad 50a + b = 1.$$

By computation, we shall find out that $a = 1/150$ and $b = 100/150$. With the help of these values of transformation constants, we shall then easily compute the numerical values of the remaining elements, namely

$$-50 \cdot 1/150 + 100/150 = 1/3$$
$$0 \cdot 1/150 + 100/150 = 2/3,$$

so that the equivalent canonical matrix will have the form

$$M_D^* : \begin{bmatrix} 0 & 2/3 \\ 1/3 & 1 \end{bmatrix}.$$

In the second case, we can choose a positive number a, arbitrary numbers b_1, b_2, multiply by the constant a all elements of the matrix M_D and add to every element in the first (second) column the constant b_1 (b_2). However, the transformation may be applied only if the conditions are equally probable.

The advantage of these transformations apparently rests on achieving an analogy with the numerical interval of probability values and on the facilitation of calculations. Yet the transformations will not enable us to establish the initial 'numerical values' of the elements of the matrix M_D, which are arbitrarily assigned to desirabilities, save the limitation that the smallest among them characterize the least desirable activity and the largest among them the most desirable one.

Because these linear transformations here explicitly refer to the preference ordering, i.e., to ordinal scales, it is evident that the invariance as to linear transformations is not a sufficient criterion of measurability of non-metric magnitudes by means of interval scales.

As we might have shown to a sufficient degree, neither the purposefulness of the chosen formalism, nor the empirical meaningfulness and operational realizability of a measurement (a much more important factor for the theory of measurement) could be justified by a purely formal determination of the uniqueness of a numerical assignment, which attempts to bypass the analysis of principal semantical problems by using sheer syntactical means. This evaluation must also be applied to the criterion of scale invariance, which is only apparently supported by the criterion of empirical operations. None of the problems related to content might be solved by formal means that have only an instrumental character. Even less could these formal means alter the borderline between that which is and which is not objectively measurable in the strict sense of the term.

8. METHODOLOGICAL PROBLEMS OF MEASUREMENT

There is a whole range of general and specific problems of a methodological character that are connected with measurement. The selection of these problems undoubtedly depends on the level of generality and the aims of the conceptual analysis and practical application of measurement. It is quite understandable that a practitioner proposing a realization of measuring experiments which are to facilitate the measurement of some magnitude in a definite specific range of numerical values, will pose and solve more methodological questions than a theoretician who has set forth the task of evaluating methodological principles applied to extraphysical measurement.

Considering the foregoing exposition of the theory of measurement and the theory of scales, in this chapter we shall first assess the function and use of the axiomatic method in the characterization of formal properties of empirical relational systems. This will be followed afterwards by a discussion of methodological problems encountered in our interpretation of empirical relations which define weak ordering and the operation of concatenation. An essentially new thematic area will consist of the problems of precision in measurement; these problems are usually analyzed in connection with the theory of measurement errors. Since this theory also deals with questions which touch upon the function of measuring apparata, the role of the observer in measuring experiments, the conditions under which measurement is practically materialized, as well as the methods used in elaborating concrete results of measurement, we cannot, within the framework of this book, devote the attention which is usually payed to it in the methodological analyses of measuring procedures. Finally, in addition, we shall investigate the questions of the meaningfulness of measurement, which must be generally solved for every kind of measurement with respect to its basic components, just as with regard to its function in the process of scientific cognition, and also the problem of so-called reliability and validity, which is laid down only in the context of extraphysical measurement.

8.1. AXIOMATIZATION OF THE SYSTEMS OF MEASUREMENT

The application of the axiomatic method in the theory of measurement has been directly connected with Hölder's axiomatization of magnitudes (Section 3.1). Such an application was first worked out by Nagel (1932) and Suppes (1951) for the systems of extensive measurement and later, under the influence of the von Neumann–Morgenstern axiomatic theory of expected utility, it was extended to various systems of non-extensive measurement as well.

Nevertheless, there exist very essential differences between the axiomatic system of Hölder and a similar system of Huntington (1902), on the one hand, and various axiomatizations in the theory of measurement, on the other.

Although Hölder's axiomatization of magnitudes aims at grasping not only the theory of abstract magnitudes but also Eudoxus' theory of proportions, the arithmetical theory of measure and the geometrical theory of distances and segments, it is an axiomatic system elaborated within mathematics and for the sake of mathematics. Therefore, from the conceptual point of view, it differs from the standard axiomatization of real numbers, as was first proposed by Hilbert, practically at the same time, only by being based on the concept of magnitude rather than on the concept of number. The ensuing terminological differences, just like the ways of concrete realization of the axiomatic construction of both these equivalent mathematical theories, are inessential. What is decisive is merely the fact that within the framework of the theoretical foundation of mathematics itself these axiomatizations have exactly the same function. When Huntington emphasized that the postulates of his axiomatic system of absolute continuous magnitudes "provide a complete logical base for a deductive mathematical theory,"[1] he was already well aware of this fact. These systems refer to idealized mathematical entities; they have a precisely delimited domain of consideration and a relatively clear semantical interpretation.

In the theory of measurement, however, axiomatization refers to various empirical relational structures, to common properties of different metric and non-metric magnitudes, to the formal characterization of experimentally realizable empirical relations and operations, or perhaps merely to a delimitation of some notion, for example, of the concept of expected utility, the definition of which is difficult to offer. Because of this specific aim, particular axioms, since they are being proposed for various empirical structures, must also be empirically testable. Hence, their validity is not guaranteed merely by the customary metalogical conditions of consistency, completeness and independence, which, at any rate, are not imposed on them in the theory of measurement, but first of all by their empirical meaningfulness as well as by their verifiability with respect to concrete properties of measured magnitudes. The adherents of the formal theory of measurement explicitly emphasize that these axioms cannot be "a too slavish imitation"[2] of the axioms of numerical systems, yet they impose this limitation only on the axiom postulating the closure of the empirical operation of concatenation. This, of course, is a serious objection against their own doctrine, for an axiomatization of empirical systems which would precisely correspond to the axiomatization of numerical systems is indeed a pivotal idea of their formal analysis of measurement. Yet their "more realistic axiomatization" of the systems of extensive measurement[3] does not refer any more to the Archimedean axiom which they need indeed, although even this axiom is not empirically "realistic", for the same reasons. Without the Archimedean axiom, the possibility of mapping empirical relational systems into a real-valued one cannot be formally justified. The weakening of the condition of additivity is again a welcome opportunity for the preparation of the ground for the axiomatization of the systems of non-extensive measurement. This self-criticism does not mean, therefore, that they would seriously consider a modification of their entire doctrine. Because of all the reasons mentioned and due to the distinct function of the axiomatic method in mathematics and in the theory of measurement, Hölder's and Suppes' axiomatizations of extensive magnitudes cannot be viewed as axiomatic systems of the same kind.

Furthermore, if we compare a construction of an axiomatic system in mathematics with the axiomatizations in the theory of measurement, we cannot indeed overlook the existence of a principal methodological difference between them. In mathematics, just as in logic or in some parts of physics, axiomatization is grounded on a certain, sufficiently rich, set of assertions which is arranged by this formal procedure into a consistent system containing, besides axioms, primitive notions, and definitions, an extensive subset of theorems, too. The usefulness of axiomatization lies in the unification of the given system of knowledge that had been previously only intuitively developed and the value of which depends upon a positive solution of the problem of completeness. Axiomatic systems are therefore consciously constructed upon semantical foundations. Without a clear conception of what it is to which this formal procedure is being applied, axiomatic systems, provided they satisfy the condition of consistency, can be viewed, at most, as a formal game that has scarcely any practical significance.

In the formal theory of measurement, axiomatization *objectively* serves only one aim: to derive only two theorems from the set of axioms postulated for some empirical structure – the representation theorem and the uniqueness theorem – and thus to demonstrate the existence of an isomorphic or homomorphic mapping between an empirical and a numerical relational system. The axiomatization of various kinds of measurement has thus only very limited deductive consequences, which are out of proportion if one considers the complexity of the formal apparatus involved as well as the real applications.

Yet it is rather the *subjective* reasons that are decisive for the introduction of the axiomatic method in the theory of measurement. The strivings for the application of measuring procedures in various domains of the social sciences – especially of the procedures of extensive measurement, in analogy with physical measurement – encounter considerable theoretical and practical obstacles. In this connection, the methodologists of extraphysical measurement usually point out the lack of well-founded theories which could be used in constructing practical schemes of measurement, as well as the lack of clarity concerning the issue of properties which, in fact, can be measured in these domains. Furthermore, they point out the absence of a clearly defined homomorphism between empirical and numerical relational structures and, in particular, the difficulties that are conditioned by the fact that we do not have at our disposal any appropriate empirical operation of concatenation.[4] However, the non-existence of measurement units, without which the measured magnitudes may not be viewed as metric ones, is to be regarded as an even more important cause of these troubles. And it is exactly the solution of this problem, as it is openly conceded by Coombs [5] in the case of psychological measurement, towards which the construction of various mathematical formalisms should contribute. To avoid assumptions leading to a general and universal measurement unit, the methodologists of extraphysical measurement lay down certain hypotheses concerning behavior which are adopted in such a manner that they correspond to some "preferred mathematical model".

Of course, nothing is solved by a production of various axiomatic systems for which there would be no need to seek some suitable empirical counterpart to the numerical operation of addition, or which would simply eliminate the problem of measurement unit. This current development in the theory of extraphysical measurement, which in principle has already been installed by Stevens' theory of scale types, appears unacceptable, even to its founder. In his criticism of the "enthusiasm for formalism as it is displayed by such authors as Suppes and Zinnes", Stevens[6] very correctly points out that their formal analysis of systems of measurement merely makes "the significance of unsolved empirical problems more obscure."

Another weighty objection may be raised against the purposefulness and legitimacy of axiomatization in the theory of measurement. We have already shown (Section 7.2) that the uniqueness theorem, strictly speaking, is not derived from the given set of axioms. The same can be demonstrated for the representation theorem as well.[7] The proof of the representation theorem is also performed on the basis of additional definitions – with the help of which such numerical expressions are introduced in the first place – that, in principle, are not derivable from the axioms postulated for empirical relational structures. This is already evident from the fact that the representation theorem correlates an empirical expression with a numerical one.

The claim that the representation theorem is strictly deductively derived from the set of empirical axioms, implies that the given axiomatic system must be deductively closed. However, this is impossible in the given case, for the axiomatization of empirical structures *ex definitione* cannot contain any primitive notion or axiom which would refer to numerical expressions. Hence, either the axiomatization in the theory of measurement violates the metalogical postulate of deductive closure, according to which the set of consequences is contained in the set of axioms, or the representation theorem is not derivable on the grounds of the accepted set of empirical axioms. In any case, this dilemma is thus a principal obstacle for the logico-methodological correctness of the axiomatic justification of the validity of the representation theorem.

Nevertheless, even if the application of the axiomatic method were correct from the methodological standpoint, i.e., even if the representation theorem had been derived only from the initial set of axioms, this would not be sufficient for regarding it as justified. If it should indeed characterize the conditions under which one can meaningfully assign definite numerical values to empirical objects, we must also show – as has already been stated – that the axioms are empirically verifiable. If we are to be consistent, then, in this case, one can neither avoid an interpretation of the basic notions of empirical structures nor a testing of an empirical relevance of the axioms.

The representatives of the formal theory of measurement[8] must, of course, also admit the necessity of a "practical empirical interpretation", yet, in fact, they concentrate rather on the properties of a numerical assignment and not on the procedures which allow this assignment. Nevertheless, in order to be able to say that this assignment is based on an isomorphism or a homomorphism between an empirical and a numerical relational system, it is obvious that the axioms for empirical structures must be chosen so that they correspond to the axioms of the numerical relational system. These axioms are thus empirical analogues of mathematical axioms, rather than assertions which would express, independently of the supposed numerical mapping, certain properties of empirical elements of measurement structures. At any rate, this is also acknowledged in this doctrine, although it is otherwise usually claimed that the axiomatization of empirical structures starts from the formal analysis of empirical relations and operations. Of course, this acknowledgment has its advantages too. If the critics of an axiomatic justification of the theorem of representation point out that not all axioms can actually be verified, this criticism can be met by discerning two subsets in the set of axioms: the subset of formal (the so-called purely structural axioms, among which one counts, for instance, the Archimedean axiom) and the subset of empirical axioms. An empirically significant interpretation and a concrete verification is then required only for the axioms of the latter subset. Yet one by means of this modification cannot refute the justified doubts concerning the meaningfulness of the use of an axiomatic method in the theory of measurement; namely, if the set of axioms also contains purely structural axioms, then this axiomatization cannot be held to be an axiomatization of empirical structures.

We can avoid such a problematic application of the axiomatic method in the theory of measurement if, for the justification of the representation theorem, we utilize a simplified procedure, founded on a direct empirical interpretation and verification of the correspondence rules. For instance, if some empirical relation is to correspond to the numerical relation '>', it is evident indeed that the assumed correlation will be appropriate only if these relations have analogical formal properties. Since it is a well-known fact in mathematics that the relation '>' is irreflexive, asymmetrical and transitive, one has to investigate whether the same, and under what limiting conditions, can be said at all about its empirical counterpart. It is therefore futile to introduce these general properties of relations by means of specific empirical axioms, if we can merely interpret the corresponding numerical axioms.

From the point of view of the methodology of deductive systems, this procedure is quite correct and logically equivalent to a converse procedure. Of course, one can verify theorems instead of axioms only under the condition of a deductive closeness of the given axiomatic system. Although this condition is not satisfied, the adherents of the formal theory of measurement cannot raise any objection against this simplified procedure; namely, they put great weight on the condition mentioned. We are left merely with an objection that these two approaches are not equally plausible from a psychological or a methodological point of view. This objection does not hold ground, for axioms are not a bit more intuitive than the representation theorem that is derived from them. Most people will comprehend more easily rules of correspondence than, for example, the Archimedean axiom and, at the same time, they will not have to follow the entire, sometimes too complex, process of derivation.

One advantage of this simplified approach lies in its prevention of a possible occurrence of formal and methodological errors. In addition, it helps to focus the theory of measurement on the elaboration of the problems of content. In this case, the overestimation of the role of axiomatization in elucidating fundamental problems of the theory of measurement cannot take place, neither can there arise illusions as to our ability to attain the measurability of magnitudes by purely formal means, without meeting, at least indirectly, the necessary conditions of metrization. Certainly, we need not be sorry for the loss of complexity and an apparent formal elegance.

8.2. EMPIRICAL RELATIONS AND OPERATIONS

The delineation of the formal properties of empirical relations and operations is a minimal requirement for the assumed correlation between the numerical and empirical components of the correspondence rules, or for the representation theorem. For the sake of their qualitatively differentiated use, one should not be satisfied merely with a general characteristic of empirical relations and operations; hence, the specification of their interpretation, both from the semantical standpoint and from the operational one, is quite inevitable.

In the first place, this additional requirement refers to the relation of empirical equality which is explicated in various contexts as the relation of coincidence, congruence, qualitative equality, sameness, indiscernibility, and indifference and, eventually, also as the relation of identity. This relation is specified for particular magnitudes by means of comparative relations of the type *just as*, for example, *just as long as, just as hard as, just preferred as*, and the like.

Even without the necessity of analyzing in detail the meaning that may be assigned to these, semantically different yet formally analogical, concepts there is no doubt that some of them are rather pseudo-empirical variants of mathematical or logical concepts. Even less can one consider identity as an empirical relation – an empirical property of the world,[9] – unless one holds the philosophical position of Parmenides. Such an understanding of identity, already refuted by Heraclitus, cannot, however, be attributed to Leibniz's logicizing objective idealism either. Leibniz's criterion of the *identity of indiscernible things* is only an ontological, not, however, an empirical, underpinning of the semantically defined concept of identity *salva veritate*. If there is any concept that has an explicitly formal nature, it is exactly this concept of identity. In the objective reality, there exist no identical objects, phenomena, or processes; nor do there exist, however, two objects which are identical merely with respect to some empirically ascertainable property.

In what sense may one then understand the notion of empirical equality? How can it be found out that two magnitudes are at least approximately equal? These problems are solved in two ways, according to the interpretation of the relationship between the concept of equality in its different explications and the concept of distinctness, which is explicated as *preceding*, (*succeeding*), *preference, greater* (*less*) *than*, and the like. According to the so-called relational theory,[10] the relation of empirical equality, in an analogy with the trichotomy of numerical relations '=', '<' and '>', is viewed as an observable relation on the same level of abstraction as both variants of the relation of distinctness, namely, the relations 'greater than' and 'less than'. According to the so-called absolute theory, based on the presupposition of the epistemological priority of empirical judgments of inequality as to the judgments of equality, equality is interpreted as a secondary relation entailed by the, observationally determinable, relations 'greater than' and 'less than'.

Nevertheless, unless we regard the empirical equality as the relation of identity, and unless we put the sensory and abstract grades of cognition into a metaphysical contradiction (in conflict with dialectical materialistic gnosiology), the difference between these alternatives rests on our view of these judgments of equality, supported by perceptions of indiscernibility, as "active" or "passive" acts of cognition.

According to the first doctrine, the formulation of judgments of equality rests upon a purposeful orientation towards the ascertainment of common characteristics of the measured objects. For example, when we compare the length of two objects, their congruence – achievable, of course, only with a certain measure of exactness – is to us a sufficient reason to utter a judgment of equality. According to the second view, equality is understood merely as a consequence of our inability to find some distinctions. If we cannot find out, or if we are not able to decide, whether there exists some difference between the investigated objects, we hold them to be indiscernible with respect to the given property. If we make decisions, for instance, between two alternatives and cannot give any reason why we should prefer one to the other, it remains but to view them as indifferent.

From the point of view of dialectical materialistic gnosiology, which understands the process of cognition of the objective reality as a process "from living perception to abstract thought, *and from this to practice*,"[11] it is the first conception that is to be accepted. The knowledge of empirical relations of equality and distinctness is a historically conditioned process, corrected by practice and contingent upon the measure of exactness with which we strive to maintain, for the purposes of the given inquiry, whether a definite property of the measured objects occurs in the same or distinct degrees. It is therefore self-evident that the discernment of the same properties, provided this is materialized by a customary observation, will be changed into the ascertainment of differences, if, for the sake of observation, we use a microscope, a telescope, or other instruments. Nonetheless, to consider only the issue of whether an "exact equality" between two measured objects is or is not directly ascertainable by sheer sense-perception, just as to view the manner of constituting judgments of equality only as a passive activity, amounts to misunderstanding the dialectics of the process of human knowledge.

It is sometimes stated in defense of this position[12] which has its origin in the empiricist theory of knowledge that, contrary to distinctness, equality cannot be considered an empirical relation exactly because the actual judgments of equality acquired by an immediate observation of indiscernible or indifferent objects, are not transitive. If empirical equality is to be a suitable counterpart of numerical equality, which is *ex definitione* a relation of the equivalence type, we must replace the non-transitive relation of empirical indiscernibility by an ideal transitive relation of equality. Therefore, this relation is no more an empirical relation in the right sense of the terms, i.e., it is not a relation independent of the higher levels of abstraction.

This argument, in fact, is not sufficient grounds to enable us to justify that the relation of equality – in contradistinction to the relation 'greater (less) than' – is not an empirical relation in the correct sense of the term; analogical idealizations refer to the relation of distinctness, too. Also for preference, succeeding, and the like, one can find cases that do not meet the property of transitivity. For the preferential ordering it is even conceded, empirically and theoretically is a more controversial, anti-intuitive presupposition of completeness – a presupposition which implies its qualitative homogeneity as well as constancy (immutability in time) and countability of thusly ordered elements.

The judgments of equality and distinctness always transcend the immediate empirical evidence by comprising more than one sense-perception. This transition from the concrete to the abstract, characterized by gradual generalizations and abstractions up to the constitution of empirical concepts of equality and distinctness and their correlations with numerical relations, is tied up with more and more conspicuous idealizations which, however, refer in the same measure to both these basic relations characterizing every ordering. Both empirical relations correspond to their numerical counterparts only approximately. It can be found out only by concrete measuring procedures whether the former are more or less congruent with the latter, and that depends on the specific character of the given magnitude, on its theoretical founding, on the required interval of measured values, and the like.

Much more complex problems are encountered at the interpretation of the empirical correlate of the numerical operation of addition. Not only are these problems conditioned by distinct epistemological positions, but they refer directly to the realization of the required homomorphism between both relational systems $\mathscr{E}$ and $\mathscr{N}$. In this case, the issue is not so much to what extent this empirical operation is idealized, but whether it exists at all and if not, whether measurement is possible in spite of it.

As is already well known, the empirical counterpart of the numerical operation of addition is assumed to be some abstractly formulated operation of concatenation (joining), for which suitable empirical specifications are then sought. If we measure, mass, for example, by means of weighing on a balance having equal arms, we may interpret this operation as the placing of two or more objects in one pan and of the corresponding number of standard weights in the other pan of the balance; if we measure length, we arrange the measured bodies in a linear way, one after the other, and the like. We can then assign to the result of this empirical operation a scale value that is equal to the addition of scale values which are assigned to unconnected objects in accord with the rule of correspondence (formula (6.3–27)). Of course, we might also imagine another interpretation, for example, the fusion of atoms or the coupling of physical bodies due to impact. However, then the resulting scale value need not be equal to the sum of individual "addends". The correspondence between the operation of addition and the operation of concatenation thus depends on the qualitative homogeneity and disjointness of the concatenated empirical objects.

Besides these "natural" specifications of the operation of concatenation, there exist still more complicated interpetations, for instance, for relative velocities in the special theory of relativity, or for trigonometric functions of angles, and in the domain of extraphysical measurement, additional "artificial" ways of concatenation are being sought too.

For relative velocity in the theory of relativity, the condition of additivity

$$\nu_3 = \nu_1 + \nu_2, \tag{8.2–1}$$

familiar from classical physics, must be expressed by a more complex formula

$$\nu_3 = \frac{\nu_1 + \nu_2}{1 + \dfrac{\nu_1 \nu_2}{c^2}}, \tag{8.2–2}$$

where c denotes the speed of light. Imagine, for example,[13] that the cosmic ship A is passing by the planet M with a relative velocity ν_1 and that the cosmic ship S, moving in the same direction, passes by the cosmic ship A with a velocity ν_2 (relative to A). What is the relative velocity ν_3 of the cosmic ship S with regard to the planet M? If the velocities ν_1 and ν_2 are small, the value of the ratio $\nu_1 \nu_2 / c^2$ in formula (8.2–2) will also be small. This value can therefore be neglected in the computation of the relative velocity ν_3 which can be calculated in accordance with the simpler formula (8.2–1). Yet, if both cosmic ships move very fast, then we must also take into account the speed of light c. The value of the velocity ν_3 will then essentially differ from the value of the simple addition of velocities ν_1 and ν_2.

For the expected utility, von Neumann and Morgenstern[14] interpret the operation of concatenation as the operation of "forming the center of gravity" of two positions with regard to the given masses that occupy these positions, or as the operation of the combination (combining) of two utilities with two alternative probabilities γ, $1 - \gamma$ $(0 < \gamma < 1)$.

Nevertheless, the combination operation, which expresses, in the von Neumann–Morgenstern doctrine, the expected property of a utility, differs distinctly not only from the operation of concatenation applied in measuring length, mass, and other physical magnitudes, but also from the operation of "forming the center of gravity", too. Since, in the combination of future events, with respect to which one makes a decision, only one alternative can actually be materialized, this operation is to be interpreted as follows: either one is to expect to get u with the probability γ, or to get v with the probability $1 - \gamma$, but never both. Hence, as to both possible events, only one can be materialized in reality. Yet in the case of "forming the center of gravity", both these components exist at the same time, for otherwise it would not make sense to talk about their common center of gravity. Therefore, the operation of combination does not produce any reality but only a conceived combination of future alternatives. This is also evident from other explications which confuse a semantical founding of this theory with its operational realizability, if the operation of combination is interpreted, for example, as a lottery, betting, or game in which one winning is achieved with a definite probability and another with a complementary probability. These interpretations provide a justification neither for the empirical character of this basic operation, nor for its relation to the operation of addition.

In contradistinction to the customary expression of the operation of concatenation for extensively measurable magnitudes

$$\phi(x \circ y) = \phi(x) + \phi(y), \qquad (8.2\text{–}3)$$

where ϕ denotes a numerical function defined over elements of an empirical set, we find in the original formulation of von Neumann–Morgenstern, namely in the formula

$$\mathrm{v}(\gamma u + (1 - \gamma)v) = \gamma \mathrm{v}(u) + (1 - \gamma)\mathrm{v}(v), \qquad (6.4\text{–}2)$$

two occurrences of the sign '+'. The contention that the occurrence of this sign on the left-hand side expresses the empirical correlate of its occurrence on the right-hand side does not explain anything. Nevertheless, neither will the interpretation of the combination operation be avoided in a purely formal manner. Many writers, in order to eliminate in advance the question of the actual difference between the empirical and the numerical interpretation of the sign '+', attempt to have it solved simply by expressing the left-hand (empirical) side of the above formula in another way, for example, $(A, p; B)$, $a\gamma b$, xpy, $\langle px, (1 - p)y\rangle$, $\mathrm{h}(x, \gamma, y)$. But, again, this symbolic convention does not settle anything.

One can find out that formula (6.4–2) does not satisfy the assumed correspondence between empirical and numerical concepts by its other syntactical and semantical characteristics: from the occurrence of the sign '=' as the chief functor and from the expression of the type v( ) on both sides of the equation. Because the sign '=' is interpreted in the von Neumann–Morgenstern system as a "true identity",[15] both expressions must have the same meaning or perhaps even the same sense. Of course, they must have the same meaning, even if we interpret them merely as an equality. But the expression of the type v( ), which is generally explicated as a definite numerical function, does not attest to the empirical character of the left-hand side.

In addition, numerical probabilities also occur in the formulation of the operation of combination. In our view, this fact cannot be explained away by regarding the probability of two future alternatives, in analogy to the "physical" concept of number, as some specific empirical magnitude.[16] One might raise similar objections against this view like those which were mentioned in analyzing the mutual relationship between a mathematical and a physical concept of magnitude (Section 3.1) and in the exposition of the notion of counting (Section 5.2).

However, we do not regard as admissible another solution of this problem that rests on the introduction of the notion of qualitative probability or on the schemata of qualitative decisions,[17] which are to facilitate the derivation of numerical probabilities. Namely, with respect to their quantitative aspects, one cannot find any difference at all between a qualitative probability $p(e)$, designating the occurrence of some empirical event e with the probability p $(0 < p < 1)$, and a numerical probability γ $(0 < \gamma < 1)$. Furthermore, nobody has managed so far to demonstrate convincingly either the meaningfulness of the notion of qualitative probability, or the derivability of numerical probabilities from the qualitative ones, or the appropriateness of the applicable operationalist procedures which would permit this manner of quantification.

In fact, the operation of combination has the meaning of the mean, probabilistically estimated value of some chance magnitude. Therefore, formula (6.4–2) either expresses an equality of two numerical expressions, or it defines a numerical value of the combination of uncertain or risk alternatives as the mean value of the components considered. This value may be computed in a way similar to the so-called mathematical hope of some chance magnitude: by multiplying every value which it can or does acquire by the probability that it will acquire the respective values, and by adding the products arising in this way.

This interpretation, however, is in conflict with the assumptions of the empirical nature of the operation of combination, for which the concatenation γu cannot be understood as a multiplication and $\gamma u + (1 - \gamma)v$ as an addition of two numerical expressions. Of course, if we consider the manner of calculating a utility function, we shall find out that the same writers who deny this for theoretical reasons, nevertheless proceed in this way. Hence, in reality, the operation of combination is merely a pseudo-empirical counterpart of the numerical operation of the addition of probabilistically estimated mean values of a utility function.

In a similar way, one also has to judge the contention [18] that extensive measurement can be applied to other extraphysical magnitudes as well – to subjective probability and risk, if we interpret the operation of concatenation in the former case as the union of disjoint events, and, in the latter case, as a convolution of probability distributions. Yet neither of these "artificial" interpretations of the operation of concatenation can be treated by us as empirical counterparts of the operation of addition.

Can we then talk about a fundamental measurement of these magnitudes in the strict sense of the term? As we have already mentioned, in the classical theory of measurement the satisfaction of the condition of additivity is considered a necessary presupposition of fundamental measurement. Although this is rejected in the so-called liberalized theory of measurement, nevertheless, besides the introduction of various systems of non-extensive measurement, a great emphasis is being put at least on the "artificial" interpretations of the operation of concatenation. This not quite consistent attitude may be explained by the fact that the theorists of extraphysical measurement derive the existence of a standard series of equally spaced elements for extensively measurable magnitudes from the empirical operation of concatenation.

Yet, in fact, the equal spacing of elements is not primarily based on empirical concatenation, just as it is not grounded on the indifference of pairs of elements in the case of an interval measurement, but on the existence of a measurement unit. This is so self-evident to the representatives of classical theory that they need not even talk about it. Nevertheless, the theorists of extraphysical measurement are silent about it for other reasons. If we can meaningfully assign cardinal numbers with which one might perform all basic arithmetical operations to elements or pairs of elements of an empirical set, we must assume that analogical empirical operations can be applied to their empirical counterparts; at least the empirical analogue of the operation of addition. Unless the scale values are allowed to be added, we cannot interpret them as cardinal numbers, but at the most, as ordinal numbers. We regard them as cardinal numbers because they reflect not only the ordering but also the size of the measured magnitudes.

The empirical additivity is a consequence, and not a presupposition, of the existence of a measurement unit. It is not a necessary condition of measurability, since in the case of temperature, for which we have at our disposal a measurement unit, we are not able to give a suitable interpretation of the operation of concatenation that could be realized just as naturally as for other physical magnitudes.

On the basis of our knowledge concerning the measurement of temperature, two antagonistic alternative solutions might perhaps be offered: either to extend the empirical operation of concatenation, or to tighten it up. As concerns temperature, in the first case we could modify this operation so that – in contradistinction to the combination of two bodies in one whole, if we measure, for example, length – the concatenation of two temperatures would be interpreted as "additivity in time" with respect to different thermal states of a body in different temporal moments. In the second case, in harmony with Campbell, one might require the satisfaction of the condition of additivity only for fundamentally measurable magnitudes. It will suffice for the derivatively or associatively measurable magnitudes to be linked to fundamentally measurable magnitudes on the ground of some theoretical law, or with the help of a certain, sufficiently justified, functional relation.

8.3. THE PRECISION OF MEASUREMENT

One of the reasons why such a great emphasis is being put on the application of quantitative methods in science, technology, and production is undoubtedly the expectation of achieving precise results. Isn't it therefore paradoxical to consider the precision of measurement or to concede at all a possibility of imprecise measurements? Yet the precision of measurement, just like the precision of any other method, conception, formulation, theory, simply any other sort of precision, is indeed not its inherent property. Hence, the question of exactness cannot be set up without regard to practical and theoretical aims and intentions, without paying due respect to concrete historico-social conditions.

It is, therefore, meaningful to speak about the precision of measurement only within the framework of some practical or theoretical context in which certain results of measurement are relatively more precise than others. Unless the relevance of an assumed result of a measurement is defined – from a practical or a theoretical point of view – with regard to the level of our knowledge, technical possibilities, and the objectives of content, we cannot say, for instance, whether the tolerance 0.1 mm is precise or not. To an instrument-maker from the seventeenth century, this would be an almost inconceivable measure of precision. Should we, for example, find out in weighing two wagons that one is heavier than the other by 1 g, it would hardly be reasonable to regard one of these results as more precise.

In comparison with counting, measurement is, in a sense, even less precise, despite the fact that in counting one uses only whole numbers, while in fundamental measurement, rational numbers are used and in a derived measurement, real numbers too. Although it is obvious that real numbers offer more precise information than do whole numbers, the precision of the results of measurement cannot be conceived in the same sense as the precision in counting elements of some set. In the first case, referring as a rule to continuous magnitudes of which it is assumed that theoretically they are infinitely divisible, one may obtain results with different measures of exactness. Yet in the second case, provided we do not miss any element of the given set, the result will be equally exact, of course, if the number of elements does not change.

The measure of precision of concretely measured values of physical and extraphysical magnitudes is also to be distinguished from the precision of values of magnitudes in the geometrical viewpoint. This immediately follows from the distinct nature of geometrical objects and real things.

The numerical value of geometrical angles is stipulated by definition in a quite exact manner. The theorem claiming that the sum of the internal angles of a triangle is equal to 180° is an analytic consequence of the postulates of Euclidean plane geometry; it is not a synthetic sentence the truth of which is to be verified by practice. Should we choose a system of postulates of a non-Euclidean geometry, this sentence would be invalid: in the hyperbolic geometry of Lobachevsky, the sum of the internal angles of a triangle is less than, and in Riemann's eliptical geometry, in turn, greater than 180°. Yet, for no triangle of a Euclidean type can it be demonstrated by measurement that the sum of its angles is indeed exactly equal to 180°, for in every measurement we shall always find out quite necessarily a smaller or greater deviation from the theoretically given value.

Besides this *defined* precision of the values of geometrical magnitudes, which is justifiable on purely theoretical grounds, for the empirical – physical and extraphysical – magnitudes, the numerical values of which are obtained only in the process of measurement, one has to distinguish two different measures of precision: the *required* and the *achievable*.

The achievable precision is objectively and historically conditioned, first of all, by the real nature of the object of measurement, the theoretical level of our knowledge and the advancement of the measuring technique. The required measure of precision depends on the purpose of the measurement in question within a certain context with regard to an achievable precision. It is, for example, the goal of astronomical measurements (through which we attempt to obtain more and more precise quantitative data that would enable us to find out new relations between thus far, less familiar phenomena) to realize the possibility of an achievable measure of exactness, though that measure is always limited, for historical reasons and objective causes. Measurements performed that depend on achievable precision are at the same time closely connected with the verification of numerical laws and theories. When our measurement respects the required measure of precision, we aim rather at the ascertainment of quantitative data, the interpretation of which has not yet been made sufficiently clear. At the same time, we also aim at the application of the results of measurement as to the confirmation or refutation of certain predictions that had been stated within the given scientific domain with regard to some relevant degree of exactness.

For practical and theoretical reasons, the measurement of a definite magnitude is always performed within some scope of exactness which is explicitly delimited or tacitly assumed. If the required measure of precision is stipulated in advance by a definite interval of admissible values, for example, 6.783 cm ± 0.001 cm, then we shall regard the numerical values 6.783 cm, 6.782 cm and 6.784 cm as sufficient, of course, in respect to this required measure of precision. Then it makes no sense to contend that one of these values is more precise or less precise than another. However, if we measure some other values, which lie outside of the assumed and required interval, for example, 6.781 cm and 6.7811 cm, then we consider the second value more precise than the first.

The required precision can, at the most, be equal to the achievable precision, yet, as a rule, is less than that. Of course, there is no difficulty in conceiving such a measure of precision that has not yet been achieved, though the achievement of such precision is not technically possible in our current period. The construction of new, more and more subtle measuring instruments and complicated measurement procedures is, therefore, one of the essential conditions for increasing the achievable precision. Is it, however, possible to make all of this more exact in an unlimited fashion? Will it be decisive for the calculation of the area of a circle if we round off the value of the number $\pi = 3.141\ 295\ 653\ \ldots$ to three, ten or one thousand decimal places? We must, however, always take into account such rounded values, for every irrational number can only be approximately numerically expressed. Then it is even more likely that by means of concrete measuring procedures, one can achieve a finite precision. The assumption of a totally unlimited measure of precision is incompatible with theoretically and practically confirmed bits of knowledge in quantum physics and information theory. Absolute precision is indeed physically impossible, for "it would mean an experiment yielding an infinite amount of information."[19]

There is here, however, another significant problem. How do we know, at any rate, that some numerical value obtained in the process of measurement has been objectively measured with an achievable measure of precision? The kernel of this question may also be formulated differently: are the values measured by empirical procedures the "real" values of the investigated magnitudes?

This problem cannot simply be solved by giving up in advance the assumption of a precise measurement and by considering the results of measurement – contingent upon the objective restrictions, the imperfection of measuring instruments and measurement procedures, just as upon the discrimination abilities of the experimenter – always more or less approximative. First of all, one should be aware that the approximative character of the empirically ascertainable numerical values might be understood in two distinct meanings. In one of them a fact is expressed that, as a consequence of the persistent motion and change of all actually existing objects and phenomena, no two results of a measurement are exactly the same. In the second sense, we understand this approximation as a dispersion of the measured values around a certain actual, "right" value. Of course, in what sense is this actual value real if in itself it cannot be reached in an empirically direct way? How can we then concede the objectivity of continuous magnitudes if we cannot fully verify them by empirical means? This discrepancy between the empirically measurable and actual values of magnitudes might be bridged over by the differentiation of different values assigned to the sizes of one and the same object of measurement.

In the first place, there is the *actual*, yet unfamiliar, value, i.e., a numerical value which cannot be obtained by means of empirical measurement procedures. Nevertheless, we assume of this value that it characterizes, independently of any manner of measurement, a definite size of the measured magnitude. For example, we might conceive it as a real number for which we do not know the last relevant place in its unfinished nonperiodic expansion.

Secondly, there are the different *measured* values, i.e., singular numerical data that are obtained in the process of measurement with the help of the most diverse measuring instruments and procedures, under optimally controllable conditions. These values always lie in a definite, more or less bounded, numerical interval.

Finally, there is the *approximate* value that is to be viewed as the most probable approximation to the actual value. This computed numerical value is equal to the theoretically estimated and statistically elaborated average of the individual measured values.

These three species of values at the same time characterize the hierarchically ordered levels of the process of measurement: the ontological, the empirical, and the theoretical. Apparently, nobody today would quite openly express his doubts concerning the necessity of distinguishing between the empirical and the theoretical level. Yet the presupposition of an ontological level, which is indispensable for the acceptance of the objective existence of strongly quantitative aspects of the measured magnitudes, will be criticized, for philosophical reasons, by the empiricistically and operationalistically oriented methodologists of measurement, as well as by the representatives of the formal theory of measurement.

Although these writers reject the objective existence of the actual value, to which we come nearer asymptotically, owing to the advancement of measuring techniques and to the intensification of our theoretical knowledge, nevertheless they tacitly assume it, at any rate, while admitting the possibility of increasing the measure of precision of empirically measurable magnitudes.

The difference between measured and actual values cannot be repudiated, because we know very well its cause. One of the possible sources is the occurrence of various errors by which every result of measurement is burdened, for this or another reason. Another source of this is the conflict between empirically measurable and theoretically presupposed values, namely, the conflict between rational and irrational numerical values. Should the empiricists consistently maintain their philosophical position, they would have to use only rational numbers and not, however, irrational numbers which are not empirically accessible. They they would have to reject the existence of continuous magnitudes; moreover, they could not use mathematical analysis based on the assumption of the continuum of real numbers. A consistent empiricism would thus lead all of physics into a pre-Newtonian, or perhaps even a pre-Leibnizian stage. Nonetheless, they do not accept this consequence and, in contradistinction to what they themselves contend, they concede the use of real numbers, even in the case of extraphysical magnitudes.

Without the assumption of the objective existence of actual values, we could not further explain why, by means of different measuring techniques and procedures, we might reach distinct approximate values, of which we, however, concede that they represent – at least for the sake of the theory

– the same value of the measured magnitude. In this case, it is of course a matter of an idealization, yet this idealization is conceptually quite inevitable if the results of measurement are to contribute to the formulation of numerical laws or to the process of making them more precise. Neither could we explain why the more perfect measuring instruments and more progressive ways of measurement, which enable us to make the already acquired, familiar values even more precise, signify such a great contribution to the advancement of each scientific discipline.

Since the most probable value of the measured magnitude, as far as it has been calculated from a sufficient number of measured values, is at each developmental stage of our cognition the most faithful reflection of the actual value, this approximation is always necessarily tied also with the effort to minimize as much as possible the influence of diverse errors of measurement.

The errors of measurement represent the measure of imprecision with which the measured values deviate from the actual value. Obviously, in this case, we are not able to establish exactly the actual error of measurement; all we can determine is its approximate value. If we exclude the so-called gross errors that are caused by an obvious carelessness of the observer or by the use of quite unsuitable measuring procedures and instruments, we can divide the errors of measurement,[20] according to their origin, into two kinds: systematic and accidental.

Systematic errors are caused by the imperfection of measuring instruments and procedures, by overlooking the changes in the environment which may influence the working of these instruments, and by the personal qualities of the observer (his discriminating abilities, the skill with which he performs measuring experiments, and the like). Systematic errors appear as simple and quite regular distortions of the data of measurement. In principle, they can be theoretically explained, revealed through the measuring experiments which have a control function, and corrected by diverse ways of measurement performed by means of other measuring instruments.

Accidental errors represent non-systematic deviations, the causes of which can be concretely conceived only with certain difficulties. Speaking quite generally, these errors are caused by the most diversified external circumstances that are not fully controllable: by a noise which accompanies any information transfer, by accidental fluctuations, for instance, by virtue of the thermal motion of molecules and electrons which occurs in all macroscopic systems. They are also the consequence of the technical limitations that, for objective reasons, do not permit the calibration of the scale of measuring instruments to be so fine that we could register, with the help of these instruments, the measured values with any arbitrary number of decimal places. In fact, the accidental errors are no errors in the right sense of the term, but rather an expression of the basic developmental lawful regularities in nature. It is therefore evident that their influence can never be fully eliminated.

An increase in the measure of precision of the approximate value of the measured magnitude, by decreasing the interval of imprecision through which, as a consequence of measurement errors, the particular measured values deviate from the mean value, is therefore limited for objective reasons. The range of possible errors might be reduced by bringing under control systematic errors and by minimizing accidental errors up to the point that it would become practically negligible with regard to a reasonably required measure of precision.

A result of a measurement is, therefore, always loaded with a certain measure of imprecision. All of this depends, however, on various factors: how this imprecision is concretely manifested; whether it converges to a definite, relatively stable value of a decisive deviation; whether the center of the probability distribution of measured values condenses around a certain numerical value; or whether the dispersion is too wide and variable. There can be no doubt that a stabilization of these values means an increase in the degree of approximation of the computed value to the actual value.

A necessary presupposition of this stabilization is the *repeatability* of measuring procedures of a definite kind with regard to the same object of measurement, under similar intrinsic and extrinsic conditions. However, in laying down the requirement of repeatability, we are very well aware that no two measurements can be performed under exactly the same circumstances and that, strictly speaking, they cannot actually refer to the same object of measurement. This knowledge is an immediate consequence of the dialectico-materialistic understanding of objective reality, anticipated by the Heraclitean *panta rhei*. In spite of its apparent unrealistic character, the repeatability of measuring experiments is nevertheless a quite indispensable condition if the results of measurement are to possess some informative value. With respect to systematic and accidental errors, one and only one measurement is, indeed, quite worthless. Their minimization is possible only if we perform a whole series of measurements. The greater and more variable this number, the more adequately we are able to ascertain the most probable value of the measured magnitude.

In order to be able to assess the relevance of the results of a measurement, at each repeated measurement we must evaluate the measurement procedures used and the intrinsic conditions under which the measurement was performed, as well as the extrinsic circumstances that could have influenced the achieved results. This control is necessary for the ascertainment of the homogeneity of these results, which have always been obtained under a slightly different objective situation. In this way we also correct the idealized assumption of repeatability under the same conditions, which, of course, had actually changed in the course of measuring experiments. A considerable measure of homogeneity may be expected in mere repetition at different temporal moments, under extrinsic conditions which are about the same as in each foregoing measurement (for instance, if we weigh some object on the same balance several times). Nevertheless, if we weigh an object on different balances, the variability of the results of measurement will, of course, be greater. In this case we must naturally take into account that different balances will work with different measures of sensitivity, so that the measured values will be burdened by other systematic errors. If we use distinct measuring instruments, for example, in the mechanical, optical, or acoustic measurement of length, the standardization of obtained numerical data will be even more complicated. Yet, what is involved in this and many other more complicated cases of a repeated measurement is essentially nothing else but an interpretation, systematization, and evaluation of the results, as in any other repetition of a certain type of experiment.

If one limits oneself to merely a few measurements, or perhaps even to a single one, this serves only to escape a difficult situation, if it is not possible, for objective reasons, to repeat this procedure more frequently. Unrepeatability of many social phenomena, if their uniqueness is emphasized above all, is thus also one of the serious obstacles for the relevance of the results of measurement in this area. Yet an explicit rejection of the requirement of repeatability for the measuring procedures in the social sciences, founded on a doubtful justification that this could "produce an impression in the interviewed persons that we do not trust their original data",[21] must be qualified as a misunderstanding of the conditions of measurability.

8.4. MEANINGFULNESS, VALIDITY AND RELIABILITY

One might consider the meaningfulness or significance of measurement from various vantage points. In the operational doctrine of measurement, the question of meaningfulness is laid down only for individual measuring procedures and techniques. In the formal theory of measurement, this question is put into a relationship with the problems of representation and uniqueness, and its solution is limited merely to the uniqueness of a numerical mapping, depending upon the invariance of scale transformations. In any of these one-sided views, it is always only one component of measurement that is being stressed and a much more principal question, whether the very assumption of measurability of some property is *as a matter of fact* meaningful – for objective, theoretical, and practical reasons – is not considered at all.

The meaningfulness of measurement in this wider sense, which we hold to be the primary one, is delimited by the objective conditions under which this procedure can be justified by the real properties of the measured objects, as well as by its function in the process of human cognition with regard to the theoretical and practical goals that we have in mind in using this form of quantification. Hence, to be able to say that a measurement procedure or a numerical mapping is meaningful, we must first assess to what extent one may understand this method, what results are expected from it, why we use it, and whether in fact we can apply it in this or another case. The solution of the question of meaningfulness of measurement on general or specific levels, is thus always a consequence of the accepted theory of measurement.

To the adherents of the wider conception of measurement, each measurement by means of a nominal scale will also be meaningful, provided that the conditions of mutually one-to-one mapping and of the empirical significance of the operation "determination of equality" are met. Therefore they will not object to the measurement of mathematical knowledge by means of a collection of questions that might be classified by using the marks "knew" and "did not know", or to the measurement of an attitude toward a sport by attendance at sporting matches or by watching them on TV; likewise, they will not object to the measurement of a degree of baldness by giving the number of hairs, of which they will moreover assume that it is measurable with the help of an ordinal scale. From the point of view of the narrower theory of measurement, however, such measurements will not be meritoriously meaningful, despite the fact that they may satisfy the condition of invariance of scale transformations just like the empirical operation of determining equality, or perhaps even the relation 'greater than'.

If we consider the notion of meaningfulness in the etymological explication "to have sense", as this notion is also commonly understood in the logical analysis of the language of science, its use falls under the province of logical semantics and not under the province of syntax, as it is claimed by the representatives of the formal theory of measurement.[22]

Unacceptable consequences of this confusion appear in the Suppes–Zinnes consideration of the conditions under which the formula

$$x + y = \alpha \qquad (8.4\text{–}1)$$

is meaningful. If it is assumed for this formula with an intended interpretation "the sum of the weight and height of each person . . . is equal to a constant",[23] that x and y are unique up to some similarity transformation, then this formula is meaningless. The reason for this, they believe, lies in the invalidity of the assumed transformation invariance. In fact, however, this is caused merely by the senselessness of the proposed interpretation.

If we transform x to kx, y to my, and α to α', where k and m are the parameters of admissible transformations, we obtain the formula

$$kx + my = \alpha', \qquad (8.4\text{–}2)$$

for which it is evident that no value of the constant α' will reduce it to an equation that would be equivalent to formula (8.4–1). It follows from this, too, that formula (8.4–1) is not meaningful under these conditions. Yet, if additional conditions are mentioned, this formula will become meaningful.

Let us consider, for example, that x and y lie on a difference scale, so that x and y will be unique up to an additive constant. If we thus substitute $x + l$ for x, $y + n$ for y, and $\alpha + l + n$ for α, where l, n are other parameters of admissible transformations, we obtain the formula

$$(x + l) + (y + n) = \alpha + l + n, \qquad (8.4\text{–}3)$$

which is equivalent to formula (8.4–1).

This formula will be meaningful even if it is, for example, assumed that y is a derived numerical assignment and that it depends in part on x. Let us assume that when we transform x to the expression kx, y will change into the expression $(ky + 2k)$. Then, in substituting α' for α, we obtain the formula

$$kx + (ky + 2k) = \alpha', \qquad (8.4\text{–}4)$$

which apparently will be equivalent to the formula (8.4–1), provided that $\alpha' = k(\alpha + 2)$.

Both writers then deduce from these two illustrative examples a remarkable conclusion indeed: " . . . whether it is meaningful to add weight and length depends not so much on the physical properties of bodies but on the uniqueness properties of the numerical assignments associated with weight and length."[24]

It will be sufficient to add to this entire consideration, the absurdity of which is perhaps sufficiently clear, the following remarks.

We regard it as evident that the addition of weights and lengths is meaningless for both physical and mathematical reasons, irrespective of the ways of transforming *ex post* the respective scale values.

If we already admit the assumption of a similarity transformability of formula (8.4–1), we see no reason why we could not correctly transform it to the formula

$$kx + ky = k\alpha, \qquad (8.4\text{–}5)$$

which is – similar to formula (8.4–3) – "certainly equivalent" to formula (8.4–1).

In comparing the transformations used, it is not at all difficult to find out that the transformations leading to formulas (8.4–3) and (8.4–4) are permitted steps of manipulating equations, as we commonly know them from elementary mathematics, yet the transformations used for the derivation of formula (8.4–2) are incorrect.

Since it follows from this syntactic treatment of meaningfulness that the meaningfulness of numerical assertions is exclusively determined by the uniqueness of numerical assignments, and not by the nature of the operations in empirical and numerical systems (as Suppes and Zinnes explicitly emphasize), they would first have to justify the meaningfulness of the property of uniqueness. We have already shown (Section 7.2) that this property merely expresses, in principle, purely tautological transformations of one numerical expression to another, moreover, without any relation to its assumed function in the theory of scale types.

Finally, we find it appropriate to note that Suppes and Zinnes, at the very end of their exposition, nonetheless concede the necessity of an additional elucidation of the problems of meaningfulness from an empirical as well as a theoretical point of view.

Both specific postulates of the theory of extraphysical measurement, scaling and testing – the postulates of validity and reliability [25] – are then linked to the thematic area of the meaningfulness of measurement and, to a considerable extent, to the problem area of precision and repeatability. Both these postulates are set forth particularly because the methodologists of extraphysical measurement are very well aware that, unlike in physical measurement, it is here often not at all clear which properties are the actual object of measurement, more precisely, the object of scaling or counting, and what conclusions can be meaningfully derived from the numerical data concerning the assumed subject matter of investigation. Since the formulation, interpretation, and application of these requirements is a subject of very vivid discussion, which so far has not reached any satisfactory and more or less congruent conclusions, in our exposition we shall limit ourselves merely to the most fundamental characteristics of these postulates.

Under *validity*, one understands roughly the validity of measuring procedures, obtained data, measuring instruments (in the given case, rather the so-called measuring tools, among which one counts – as we have already mentioned – various scales, questionnaires, interviews), simply all components of measurement and scaling. It is then asserted of a certain measurement (measuring procedure, tool, result, and the like) in an apparent tautological formulation, that it is valid exactly when it indeed measures what we think (expect) it does. Validity is also regarded as the measure of deviation from the project of a measuring experiment, from the assumed object or scaling, the chosen measurement procedure or technique, the measuring tool used, and the like.

In connection with the problem area of precision, it is also said about a certain measurement or scaling which is valid and the results of which are practically devoid of systematic errors, that it is *exact*. Although both these conditions of exactness are mutually independent – for the former refers to the project of a measuring experiment, while the latter to its performance – nevertheless, with regard to the expected result, validity is a necessary condition of exactness. It may happen, for example, that a certain intelligence test will be practically applied without systematic errors, yet it will not refer to intelligence but to memory. The achieved results are then, with respect to the goal of the inquiry, neither valid nor exact.

Under *reliability*, one understands roughly the validity of the quantitative, or perhaps merely qualitative, results, depending upon the objective trustworthiness of the measuring (scaling) procedures, techniques or tools, as well as upon the subjective trustworthiness of the respondents and the experimenter. Reliability is also held to be the measure of stability of the measuring tools with which approximately the same results might be obtainable under repeated measurements or tests. Reliability is, therefore, put into an immediate relationship with the possible errors of measurement, yet, in contradistinction to validity, primarily with the accidental errors. The trustworthiness of the results of measurement, scaling, or testing will be greater, the greater the variability and range of the interval of possible deviations from its assumed value.

In connection with the problem area of precision, it is also said of a trustworthy measurement (scaling, testing), for which one might minimize the number of accidental errors, that it is *precise*. The measure of precision, which is ascertained by the evaluation of the reliability of individual components of measurement (scaling, testing), is also a criterion for the assessment of the suitability of the chosen procedure.

However, in what measure are these postulates actually satisfiable in concrete cases? Our doubts concerning the applicability of the postulate of reliability might be justified by the difficulties connected with the repeated measurement of social phenomena. The problematic nature of the postulate of validity lies, in turn, in the theoretical and methodological lack of clarity with regard to many problems of extraphysical measurement, and particularly in the striving of many methodologists and practitioners for the broadest possible utilization of quantitative methods, without having objectively sufficient reasons to support this.

9. PHILOSOPHICAL PROBLEMS OF MEASUREMENT

The elucidation of basic notions, the investigation of empirical and mathematical components of measurement, the analysis of its formal characteristics, as well as the practical application of measuring procedures in science, technology, and production – all of this cannot be merely a matter of the general theory of measurement itself or of the methodology of empirical sciences, as it is held by the prevalent majority of scientistically oriented methodologists and theoreticians of measurement. To lay down and to solve any essential problem referring to the conceptual, methodological, and theoretical aspects of measurement means, in principle, to pose a philosophical problem too, irrespective of the question of whether this is openly admitted or silently passed over. At any rate, everybody could have acquired sufficient proof of this from the entire foregoing exposition in which we have critically dealt with operationalism, neopositivism, and the formalistic philosophy of mathematics.

In order to explain the essence and function of measurement in the process of our cognition, we cannot limit ourselves merely to a critical analysis of philosophical presuppositions and consequences of this or any other partial problem, but we must set forth a principal question of the philosophical foundations of measurement. The philosophical founding of measurement will also enable us to assess the actual possibilities and limitations of measurement.

9.1. MATERIALIST FOUNDATIONS OF MEASUREMENT

Positivistic scientism does not, in principle, formulate the question of philosophical foundations of measurement in its explicit form. And as much as it is at all concerned with this question, positivistic scientism attempts to maintain the appearance of a "neutral" attitude. Nonetheless, even in the discussion between operationalism and its critics, among whom we already find philosophically very concerned methodologists,[1] the core of the conflict lies rather in the area of methodological considerations concerning the mutual relationship between the theoretical and operational aspects of measurement or in the logical analysis of the so-called operational definitions. Of course, in all these cases, the neutral attitude towards the philosophical foundation of measurement is nothing but an illusion or self-deception. In any event, irrespective of whether this problem is posed or avoided, every doctrine of measurement must directly or indirectly react to it.

The problem of the philosophical foundations of measurement may be specified by the following questions: Does measurement have any ontologico-gnosiological support in objective reality, or does it not? Do there exist quantitative aspects of the measured objects, phenomena, and processes independent of the measuring procedures, or do there not? Are different magnitudes historically conditioned reflections of the dialectical unity between qualitative and quantitative aspects of real objects, or are they merely a consequence of a measurement or an expression of conceptual conventions?

From the standpoint of the materialist understanding of the foundations of measurement, it is always the first alternative that is to be accepted. An affirmative answer to the second alternative is characteristic of the various forms of subjective idealism.

In its extreme form, this idealist position is advocated by so-called *homocentric operationalism* represented particularly by Dingle.[2] According to homocentric operationalism, any specific operation, the result of which leads to some numerical datum, is a measurement. Measurement itself is taken as merely referring to a performed operation and might be described without any consideration of a "hypothetical 'something' that is to be measured". No differentiation can be carried out between the measured object and the means of measurement, and for this reason one is to reject a belief in objectively measurable properties of bodies. Likewise, one has also to reject the assumption that quantities have some size, for this would mean that they exist independently of the measuring operations. To homocentric operationalism, which quite openly proclaims its antitheoretic, voluntaristic, and extremely subjectivist character, such a view is unacceptable. This view leads to the objective existence of magnitudes and, thus, to the acceptance of an objective reality independent of our private experience. In the intentions of this extreme variant of a "pure" empiricism, the experience itself is merely defined by experience. If the experience, or more precisely, the personal experience of an individual, is the point of departure as well as the result of the cognitive process, the objective existence of magnitudes is not admissible either. Although the operationalists concede that various measuring operations exist in nature, these operations are understood by them in a purely subjective manner. The process of measurement is precisely specified only with respect to those details which are indispensable for safeguarding the utility of the obtained results for the sake of correlating the elements of our experience.

We encounter this doctrine (which in Dingle only appears in a more explicit form) in Bridgman[3] too, though his *methodological operationalism* concentrates on the problems of operational analysis, delimiting operational criteria for the definitions of concepts, rather than on philosophical considerations concerning the nature of measurement. In harmony with his general thesis which states that a concept is synonymous with the corresponding set of operations, Bridgman repudiates the objective character of magnitudes and their theoretical connection with the entire conceptual system of the given scientific domain; he reduces them merely to the applied measurement procedures. Bridgman makes use of a terminological convention, in order to prepare in advance a suitable ground for this reduction, as well as to avoid the necessity of expressing his view concerning a possible question of what it actually is to which measurement procedures refer (operationalism always finds this question very unpleasant). Instead of talking, for example, about length, Bridgman proposes to speak rather of "measuring length" or, even better, about a "length measurement". Because in this way this magnitude will not be linguistically expressed by a substantive, it cannot be viewed as "something". Then, too, it will not be necessary to answer the question of whether length is something more than that what we measure, if we perform the measurement of length according to a prescription.

This terminological "convention" does not require any additional comment. Yet we must nevertheless ask under what conditions one might ever find out, with respect to some operation, whether it is indeed appropriate as to the measurement in question, unambiguously defined, and immediately applicable. Bridgman did not, and could not, find any satisfactory answer to this fundamental question. The discrimination of the so-called "good" and "bad" operations is vague, just as is his pragmatic explication of a "good" operation as an operation which survives in practice, while it survives exactly when it is simple and suitable. Without establishing some criteria under which an operation is admissible, any decision concerning its applicability is completely arbitrary. The arbitrariness, conventionality, and subjectivity of the choice must also, at any rate, be manifested in the specification of particular operations, since, from the operationalist viewpoint, the applied procedures are always derived only from the personal experience of the experimenter.

Since a whole battery of serious critical objections against operationalism has been raised – not only from the camp of Marxist philosophers but also by many representatives of scientism[4] – it is perhaps not necessary to demonstrate in detail that the philosophy of operationalism cannot be a suitable foundation for a fertile theory of measurement. If measurement depends exclusively on measuring procedures and if the object of measurement does not exist objectively, we do not have at our disposal any intersubjective criterion of measurability. Should we accept this position, which is consciously based on the rejection of the ontologically binding nature of the theory of measurement, we would also have to admit that the scope of measurement might be arbitrarily extended, almost without any reasonable restriction.

For operationalism, there exists no historically conditioned boundary between that which is measurable and that which is not. Because measurement procedures are produced by each individual experimenter, the decision whether, in the given case, we are to deal with measurement or not, rests purely on the inventiveness or skillfulness of the experimenter when he proposes various ways of operationalization. Thus, from the operationalist point of view, this subjective ability is the only relevant criterion of measurability. Then we can regard every operation in which we assign numbers or numerals as a measurement procedure. Quite openly, Dingle[5] concedes this unacceptable consequence. According to his view, even the throwing of coins is a special case of measurement, for this operation, too, can be described in all empirically relevant details. Therefore, there cannot even exist any essential difference between this game and a measurement of some magnitude, say, length.

A set of operations cannot be delimited purely empirically, without the knowledge of numerical laws, without dependence on a theory. Measuring operations cannot, therefore, be the sole criterion for the definition of a magnitude, even less of a metric concept. The familiar operationalist thesis, according to which for each distinct set of operations one is to introduce some specific concept, conflicts with the primarily theoretical character of every metric concept. A differentiation of various concepts, contingent upon the variability of the set of operational procedures used, would mean a disintegration of scientific theories; it would also make impossible the use of numerical laws. If the operational doctrine of measurement were consistently realized, it would have to reject all theoretical aspects that are not immediately empirically manifested; nor could this doctrine admit irrational numerical values. Of course, measurement would thus be reduced merely to a fundamental measurement of magnitudes, with the values in a definite, specifically defined interval of rational numbers.

The claim that only the applied measuring procedures are decisive in the process of measurement leads, however, to even more absurd consequences. If, with regard to the measuring instrument, a physical magnitude manifests itself only as some number "obtained by some operation which belongs to a certain class of operations",[6] an admissible operation will also be the operation of "reading the scale of a measuring instrument", or the operation of "ascertaining the size of a deflection of a pointer on the scale of a measuring instrument." Yet, from the standpoint of operationalism, numerical operations obtained by this operation and applied on different measuring instruments, define one and the same magnitude. Two conflicting consequences therefore follow from the operational doctrine: a splitting of a unified concept into distinct ones, and the merging of very heterogeneous concepts into one, and only one, concept.

The aforementioned criticism of homocentric and methodological operationalism also refers to some degree to a relatively moderate position of Ellis who, among all the positivist theoreticians, devoted by far the most attention to the philosophical questions of measurement. The influence of operationalism is exhibited particularly in his conception of scales of measurement as well as in his interpretation of the concept of magnitude. Ellis[7] rejects the view that "quantities[8] have a kind of primary ontological status", and denies the "assumption that bodies have properties which have magnitudes". Because Ellis does not admit the ontological priority of the quantitative aspects of objectively existing things, phenomena, and processes, he assumes that magnitudes [quantities] cannot be a conceptual reflection of these aspects. Nevertheless, Ellis does not identify magnitudes with measuring procedures but with a "generated ordering": "the existence of a quantity is entailed by the existence of a set of linear ordering relationships." A magnitude is then measurable if different measuring procedures can "generate the same order amongst the same particulars under the same conditions".

Yet, this solution cannot be considered a viable philosophical point of departure for the justification of the theory and practice of measurement. In the first place, Ellis' doctrine is loaded by the difficulties of explaining what is to be understood by the expressions "the same particulars" and "the same conditions". Even more problematic, however, is the issue of the practical realizability of this "sameness"; if the objective ground of the measured magnitudes is rejected, then it is not at all clear what it is to which this "sameness" refers. Besides, it is evident that a stipulation of the same ordering is not in itself a sufficient criterion for the definition of qualitatively distinct magnitudes: all non-metric magnitudes, just like all metric ones satisfy, irrespective of their qualitative specificity, the same properties of ordering.

Furthermore, the relation of ordering merely defines the topological conditions of metrization. This relation is therefore a necessary, and not, however, a sufficient, condition of measurement in the strict sense of the term. An ordered sequence of objects which have some common qualitative characteristic, just like a quasi-series of scale values corresponding to them, cannot in itself be the fundamental, further unanalyzed, basis of measurement. The ascertainment that some relation fulfills the properties of an ordering, is already a consequence of the practical and theoretical activity of people – a result of the historical process of the cognition of quantitative aspects of the objectively existing objects or properties which really exist in various degrees. No theoretician or practitioner, who puts as his aim of inquiry a performance of measuring experiments, will begin his work by investigating whether it can be said about some property common to a definite class of objects that it fulfills the conditions of topologization. The point of departure of his activity is always a certain content analysis of the investigated objects, of which he assumes, on the ground of some working hypothesis – correctly or incorrectly – that they will have definite weak or strong quantitative aspects. Only then will he be able to say whether these objects may be ordered with respect to the property in question, or not.

From the standpoint of the materialist doctrine of measurement, ordering relations, as well as measuring operations, are based on the assumption of the existence of quantitative aspects of real objects. The measured magnitudes are definitely neither a product of measuring procedures nor a consequence of the ascertained ordering relations, but a reflection of the dialectical unity of qualitative and quantitative aspects of real objects, phenomena, and processes. The assignment of numbers to empirical objects, whether these are obtained by a fundamental, derived or associative measurement, is empirically significant and theoretically warranted only if it refers directly, or at least indirectly, to their quantitative aspects. Magnitudes are objectivized by a measurement and not, however, constituted by it. An assignment of numbers to objects, phenomena, and processes, "according to rules", must fully respect the objectively existing distinction between their qualitative and quantitative characteristics. By measurement, we cannot arbitrarily change the ontological nature of the investigated property – reduce qualities into quantities. Neither can this be achieved by arbitrary conventions or theoretical constructions based on the requirement of simplicity, as is believed by the instrumentalists and conventionalists, nor by means of various mathematical models, as is presumed by the representatives of the formal theory of measurement.

The application of measuring operations, just like the stipulation of ordering relations, are significantly realizable only up to the point that objective reasons exist for this. Without reference to objective reality, all quantitative assertions about the results of measuring experiments would be merely assertions concerning measuring operations or ordering relations. They would be only statements referring to subjective experience and not, however, assertions dealing with intersubjectively verifiable facts.

The history of science provides enough proof to the effect that measuring procedures developed in the process of a practical appropriation of the world by man according to the quantitative aspects of real objects. In its essence, from the very origin of its existence, measurement is a complex of mutually conditioned empirical and numerical components, operational procedures, and theoretical doctrines. It also follows from this mutual connectedness of all its essential characteristics, that it would not be right to deny the influence of measuring procedures. Nor, if we are to deal with magnitudes which are beyond any doubt immediately dependent on objectively quantitative aspects, can we draw a conclusion from the criticism of operationalism that these magnitudes would be absolutely independent of the measuring procedures used. Such a standpoint would be erroneous, for matter-of-fact reasons as well as for philosophical reasons. Nevertheless, the manner in which we measure a magnitude or the choice of measuring operations and measuring instruments are not at all arbitrary, but always subordinated to the real properties of the measured objects, to the conceptual and theoretical system of the scientific domain in question, as well as to the needs of practice.

The ontologically binding nature of the procedures, conceptions, and objectively justifiable conventions in the theory of measurement also fully refers to the scales of measurement and to the choice of their origin and measurement unit. We have already pointed this out in our criticism of the overestimation of the difference between the so-called arbitrary and the so-called natural origin of a scale (Section 4.2), as well as the conventionalist interpretation of the notion of a measurement unit (Section 3.2). At any rate, the fact that, in the case of extraphysical measurement, we do not have at our disposal an objectively reproducible and significantly interpretable measurement unit, is the most convincing argument against the conventionalist view of a measurement, as well as against the anti-ontological position of operationalism, instrumentalism, and neopositivism.

Of course, the priority of objectively existing quantitative aspects as to the measuring operations (the dependence of conventions at the choice of measurement units upon the real properties of the measured magnitudes), should not be viewed metaphysically, in the sense of contemplative materialism or naive realism. Contemplative materialism, which does not adequately value the active role of man in the process of cognition, cannot therefore provide an adequate philosophical foundation for measurement. With respect to operationalism, its one-sidedness is demonstrated only in the opposite direction. Operationalism unjustifiably extends the concept of measurement by emphasizing merely its conventional side. In turn, contemplative materialism narrows down this concept, for it regards magnitudes only as a passive reflection of the quantitative aspects of real objects. Concisely speaking, operationalism is based on a "bad" ontology, contemplative materialism in turn on a "bad" theory of knowledge. Besides, neither operationalism nor contemplative materialism – as a consequence of their ahistoric position – are able to explain for the theory of measurement so important a problem the transition from classificatory and comparative notions to metric ones. Because they ignore the historical conditions under which measurement advanced in various scientific domains, they cannot elucidate the objective circumstances under which measurement is indeed possible.

A philosophical foundation of measurement which would rest on the principles of Marxist-Leninist philosophy, as well as the criticism of the operational doctrine of measurement, have become urgent desiderata, particularly in light of a very extended application of measuring procedures in various domains of the social sciences.

In physics, which has always been developed on materialist foundations – owing to the subject-matter of its investigation as well as to its applications in technological practice – the danger of an operationalist misinterpretation has not been so acute. Since physics had already reached a considerable theoretical level in Newton's time, operationalism could not succeed more expressively because of its antitheoretic orientation. Nor has this doctrine set stronger roots in experimental physics which is overlapping more and more closely with theoretical physics. Therefore, the awakening from a transient operationalist intoxication here has been much easier and more consistent.

The same, however, cannot be said about psychology, which, in its endeavor to overcome its traditional, speculative, rather than empirical way of inquiry, has welcomed the methodological postulates of operationalism. It appeared, especially to psychologists who have taken aim at the practice of testing, that the operational approach to the theory of measurement obliterates any essential differences between measurement in physics and psychology and, hence, that only operational psychology is indeed a scientific psychology.[9]

Of course, it must be conceded that, at the beginning, the operational approach in psychology indeed signified a step forward, in comparison with the traditionally cultivated "spiritual" psychology. However, as soon as the need was felt to incorporate operationally obtained data into broader theoretical connections and to seek suitable theoretical justifications for the testing and measuring procedures, the methodologically immature and philosophically erroneous postulates of operationalism began to manifest themselves in a very unpleasant way. Yet, neither the construction of various, often very complicated, testing procedures, for which it was very difficult to give an objectively satisfactory interpretation and theoretical justification, nor the effort to reach exactness by introducing quantitative methods, nor the overestimation of the possibilities of extraphysical measurement, have led to desirable results. Attempting a consistently empirical approach, psychologists relied on the universality of measurement and were aware that for each extraphysical measurement, the respective procedures had already been theoretically justified in a sufficient way. The emphasis put on operational definitions and the operationalization of only very vaguely expressed hypotheses have lured their attention away from more systematic theoretical analyses referring to both the problems of extraphysical measurement and to the proper area of inquiry. Naturally, these problems cannot be solved with the help of methodological and philosophical doctrines of operationalism, which had already been wrecked in physics.

Of course, later the methodologists of extraphysical measurement also separated themselves from operationalism, for they could learn from their own experience that reliance merely on measuring operations is far too insufficient to warrant an actual validity of the numerical results that have been obtained by numbering, scaling, measuring, or counting, simply by any "numerical quantification" of the social phenomena.

The road which Stevens took with his conception of the invariance of a scale form, and in whose footsteps there followed, even more consistently, Suppes, Zinnes and many other methodologists of behavioral and social measurement by deriving the theorems of representation and uniqueness from the axioms of relational structures, leads only from one extreme to another. The departure from the subjective empiricism of the operational doctrine of measurement in favor of the objective formalism of a purely mathematically developed theory of measurement, which from a philosophical point of view pays tribute to the so-called model platonism, does not contribute towards the theoretical or philosophical foundation of measurement.

As before, this development is conditioned by the lack of a critical attitude toward the problems of extraphysical measurement. Yet, there do not exist the same objective conditions and presuppositions for this kind of measurement, owing to the specific nature of the social scientific disciplines, as contrasted with the natural and technological sciences, particularly physics. Thus, this development again only confirms the legitimacy of the materialist foundations of the theory of measurement.

9.2. THE POSSIBILITIES AND LIMITS OF MEASUREMENT

In the above-mentioned outline of the materialist foundation of measurement, we have followed the philosophical questions already discussed in connection with the previous exposition of basic notions, and theoretical and methodological problems of measurement. As we have frequently stressed, from this outline it follows with great force that measurement is limited by objective reasons, in particular by the existence of quantitative aspects of real objects, phenomena, and processes.

This objective foundation of measurability, then, conditions the properties of metrization, the character of numerical mapping which depends on the existence of measurement units, as well as the manner in which this empirico-mathematical procedure can be practically realized in various areas, with regard to their specific subject-matter of investigation. The scope of measurability is further historically conditioned by the attained level of theoretical knowledge, practical requirements and technological possibilities tied up with the construction of suitable measuring apparata. From the methodological function of measurement, there also follows the necessary requirement of an appropriate theoretical interpretation of obtained numerical data, also facilitating at the same time an extrapolation or interpolation of those numerical values which are not directly obtainable by the measuring procedures. Without reference to a certain theory, scientific hypothesis, or at least some hypothesis *ad hoc*, measurement itself can produce only results concerning which we cannot even say to what it is they refer. Measurement makes sense only if it produces a basis for a formulation of numerical laws – if it contributes towards a confirmation, verification, or falsification of some theory or hypothesis.

One must always fully respect these objective limitations, as well as the necessary ties between measurement and theory, between the aim of measuring experiments and practical needs, if this procedure is to lead to theoretically justifiable and practically applicable results. There is, however, an additional important consequence that follows from the materialist foundation of measurement: *some properties of real objects are not measurable in principle.*

Of course, the philosophical question of measurability, which in this or another form must be set forth by every theoretician of measurement, might be answered in the intentions of subjective idealism, too. This is the how Carnap[10] quite explicitly solves this problem. His approach is based on a purely phenomenalist understanding of objective reality, on a nominalist reduction of concepts to mere linguistic expressions, as well as on an idealist interpretation of mathematics. In accordance with the operationalist doctrine of measurement (which, however, he does not accept fully, due to its antitheoretic orientation), Carnap considers every ordering (inasmuch as it permits a mutual comparison of different objects with respect to some common property) to be a sufficient precondition of measurability. Anything else is then only a matter of suitably proposed procedures, with the help of which one may assign numbers to phenomena in a "useful way."[11] From the assertion that "quantitative concepts . . . are part of our language, not part of nature", which misinterprets the dialectics of the process of knowledge, and from the assertion that "the phenomena themselves exhibit only qualities that we observe", which is an expression of a subjectively idealist interpretation of the notion of experience, Carnap infers that "it is *we* who assign numbers to nature." In assigning numbers to objects of our experience, we need not lay down at all the question of "whether they are the 'right' numbers." It is simply enough to devise certain rules that specify the procedures facilitating this assignment. On the grounds of this explanation of the philosophical questions of measurability, Carnap then reaches the conclusion that "from this point of view, nothing is in principle unmeasurable."

Nevertheless, Carnap's justification of the doctrine of the entirely unlimited possibility of measurement, which cannot be accepted on its merits just as one cannot accept the doctrine itself, has great worth for us, for it reveals very openly the idealist roots of this position. In addition, this justification provides a very conclusive argument against the conception of measurement in the wider sense of the term. If we could assign "numbers to nature" with the help of rules in the sense of Stevens' definition of measurement (which Carnap consciously or unconsciously accepts) without being compelled at the same time to set up the question of whether this is also warranted for objective reasons, then we would have to take for a measurement any manipulation with numbers, or perhaps even with numerals, insofar as it would refer to empirical objects. In our exposition we have already sufficiently demonstrated the principal unacceptability of such an "overflown" doctrine of measurement; it is therefore unnecessary to repeat our critical objections once again.

There can be no doubt that measurement is one of the basic scientific procedures, the significance of which grows constantly, particularly in the period of the scientific-technological revolution. However, this fact must in no case seduce us toward an overestimation of its possibilities. A glaring prospect of numerical calculations, the attractiveness of the formal approach and striving after an increased measure of precision must never predominate over the real conditions under which measurement is in fact realizable. A sober estimation of the conditions of measurability contributes much more towards the advancement of science, technology, and production than the subjective view that measurement is essentially possible without any objective limitation.

Obviously, these limitations will be much greater in the social sciences than in the natural and technological sciences. This does not mean, of course, that we would have to take a sceptical position against extraphysical measurement in the proper sense of the term. In the social sciences, we are still far from utilizing all the possibilities of obtaining quantitative data by counting and of developing different ways of scaling. One can also expect very positive results from the associative measurement which has not yet received the attention that it rightfully deserves. Of course, it is not sufficient to concentrate our attention exclusively on the operational and formal aspects of these quantitative methods and to overlook, at the same time, much more important problems of a philosophical, theoretical and methodological nature.

Mathematization in science has only an instrumental character. The usefulness of this approach depends on its thoughtful usage, on the elucidation of its theoretical and methodological aspects, as well as on its appropriate specification with regard to a concrete object of investigation. The application of quantitative methods, mathematical models, axiomatization or formalization cannot be a substitute for the solution of the fundamental theoretical conceptions, for the advancement of specific methods of the scientific domain in question, or for the process of making the concepts employed more precise. Exactness cannot be reached simply by expressing imprecise and obscure conceptions in the language of mathematics.

Quantitative methods, mathematical models, axiomatization or formalization, inasmuch as these are being applied by bourgeois methodologists under the influence of neopositivism and a formalistic philosophy of mathematics, irrespective of their objective presuppositions and social-historical conditioning, cannot fulfill the function which is expected from them. Mathematization in science cannot become a new, or perhaps even the exclusive method of inquiry; it can never replace the dialectico-and historico-materialist approach to a scientifically founded analysis of the given phenomenon or object of investigation; it is always only a means helping to objectivize the results of inquiries in the social-scientific domain in question.

NOTES

CHAPTER 1

[1] Cf. Farrington (1969); Bernal (1971); Struik (1967). [These standard references to the Bibliography will be used uniformly in all Chapters – all translator's notes will be put into brackets.]
[2] Filkorn (1953), p. 58.
[3] In our [i.e., Czechoslovak] system of measurement it is one kilogram. Our system is established by the binding Czechoslovak State Norm 01 1300 from 3 January 1963 (effective as of 1 August 1974), in accordance with the International System of Units (Système International d'Unités) SI.
[4] Engels (1964), p. 256.
[5] For the purposes of the theory of measurement, we consider it more suitable to use the terms *classificational, topological*, and *metrical*, rather than the synonymous terms *qualitative, comparative* and *quantitative*, which can be encountered in the works devoted to the construction of the language of science.
[6] Cf. Filkorn (1963), pp. 69, 193.
[7] Cf. Weyl (1927), p. 100, and Weyl (1949), p. 139.
[8] Bernoulli (1738/1954).
[9] von Neumann and Morgenstern (1944/1953), p. 16.
[10] Fechner (1860).
[11] Cf. Campbell (1940).
[12] *Helmholtz (1887/1971). [Asterisks denote references added by the translator.]
[13] A unification of this sort is the goal of a general conception of metronomics, elaborated on the basis of physical and technical measurement by the research workers of the Institute for Measurement and Measurement Techniques of the Slovak Academy of Science. Cf. especially essays of I. Staríček in *Acta Metronomica.*

CHAPTER 2

[1] Brož (1967), p. 17f.
[2] Kockelmans (1966), p. 115. Cf. likewise Margenau (1950), p. 369f.
[3] Cf. similarly, e.g., Lindsay (1938), p. 456, " . . . ultimately all measurement reduces to attaching a number to the coincidence of a pointer and a scale."
[4] Lamser (1966), p. 150.
[5] Goude (1962), p. 14.
[6] Cf. similarly Lazarsfeld (1959), p. 108, " . . . the attribution of properties is interchangeably called description, classification, or measurement."
[7] Cf. likewise Ackoff *et al.* (1962), p. 177: "Measurement . . . is a procedure by which we obtain symbols which can be used to represent the concept to be defined."

[8] Wiggins (1968), p. 417.
[9] Because of the step-by-step nature of our analysis, it is necessary to give references to later and previous sections of the present work.
[10] Cf. e.g., Pfanzagl *et al.* (1968), p. 31.
[11] Cf. e.g., Stevens (1959), p. 22; Ellis (1966), p. 156f. [The term *numerosity* is used by Stevens (1951), p. 22. A bit cumbersome, the term *numerousness* is also used in the literature – tr.]
[12] Cf. successively Campbell (1920/1957), p. 267; (1928), p. 1; (1940), p. 340.
[13] Cf. Campbell (1920/1957), pp. 268f., 295f.
[14] Campbell (1920/1957), p. 305.
[15] Campbell (1920/1957), p. 313.
[16] Cf. Stevens (1946/1960), p. 142; (1951), pp. 1, 22; (1959), pp. 19, 25.
[17] Stevens (1946/1960), p. 148.
[18] Stevens (1946/1960), p. 145.
[19] Adams (1966), p. 125. [Adams criticizes this theory – tr.]
[20] Cf., e.g., Ujemov (1970), p. 68; Berka (1974a), p. 225f.
[21] Cf., e.g., Essler (1971), II, p. 64f.
[22] Bunge (1967), II, p. 220.
[23] Stevens (1959), p. 25.
[24] Cf. Goude (1962), p. 11.
[25] Peak (1953), p. 250; Torgerson (1953), p. 12; Ross (1964), p. 133.
[26] Stevens (1946/1960), p. 144; Coombs (1953), p. 473; Suppes and Zinnes (1963), p. 23. This very wide conception of measurement is sometimes also called "postulated measurement". Cf. Torgerson (1958), p. 22; Pfanzagl *et al.* (1968), p. 33.

CHAPTER 3

[1] *Met.* V c. 13 p. 1020a [tr. R. Hope, 1960, The Univ. of Michigan Press, Ann Arbor, Mich., p. 107 – tr.] Cf. also *Cat.* c. 6.
[2] *[English translation in *Essays on the Theory of Numbers*, tr. W. Woodruff Beman, 1963, New York – tr.]
[3] Archimedes, *On the Sphere and Cylinder*, quoted after Struik (1967), p. 46, n. 3.
[4] Cf. Stolz and Gmeiner (1900), p. 13.
[5] Bolzano (1950), p. 91f. Cf. p. 96: "Let us choose any two abstract quantities, say 5 and 12." [It may be worthwhile to quote a more extensive passage: " . . . not only are all numbers also quantities, but there are far more quantities in existence than there are numbers: because the fractions . . . and likewise the so-called *irrational* expressions $\sqrt{2}$, . . . denote quantities . . . " – tr.]
[6] Lambert (1771): for more detail cf. Berka (1973a), p. 231f.
[7] Hilbert (1900), pp. 180–194; cf. also Hilbert (1930).
[8] Stolz and Gmeiner (1900), p. 7f.
[9] Huntington (1902), p. 266f; Hölder (1901), p. 1.
[10] For example, in expressions of the type "that something, e.g., 'length' is a magnitude" and "two entities of a certain kind have the same magnitude." [These expressions are quoted in English, not in Czech – tr.]
[11] Campbell (1920/1957), p. 238; Campbell (1928), p. 14.
[12] Cf. Hempel (1952), p. 55; Bunge (1967) II, p. 198.

[13] Hempel (1952), p. 85, note 61.
[14] Cf., e.g., Russell (1903/1964), p. 159; Pap (1962), p. 126.
[15] Cf., e.g., Focken (1953), p. 1.
[16] Cf. Menger (1959); cf. also Körner (1970), p. 176f.
[17] Cf., e.g., Ujemov (1963).
[18] Lenin (1961), p. 93.
[19] A more detailed criticism is given, for instance, by Čížek (1974).
[20] Cf., e.g., Binko (1968), p. 17; Stille (1961), p. 10f; Flaschner (1952), p. 141; Alberti (1957), p. 7.
[21] Cf. Sedov (1955), Brož (1967), Binko (1968), Bridgman (1922/1931), Buckingham (1924), Causey (1969), Focken (1953), Krantz *et al.* (1971), etc.
[22] Krantz *et al.* (1971), p. 460.
[23] Cf., e.g., Bunge (1967) II, p. 226.
[24] Cf., e.g., O'Rahilly (1938), p. 748; Jeffreys (1943), p. 838.
[25] Apparently, without extensive use of the theoretical knowledge of physics, it will not help much if we try to clarify the definition of the concept of meter by mentioning that krypton 86 is an isotope of krypton with a nucleon number 86, that $2p_{10}$ and $5d_5$ are spectroscopic designations of the respective energy levels of the atom, and that the respective radiation has a red color and the wave length $\lambda \doteq 605,780\ 2$ nm (nanometers, i.e., as to the rank 10^{-9} m).
[26] Engels (1964), p. 252.
[27] For example, to the measurement system MKSA – *meter, kilogram, second, ampere* – which is contained in the SI system. [For a thorough explanation, see Page and Vigoureux (1975) – tr.]
[28] Cf., for example, Flaschner (1952), p. 146f.
[29] The dimensional formula of length is the expression $[1] = L$.
[30] Formerly (until 1 August 1974) 1 kilopond (kp) was admitted as a subsidiary unit of force. 1 kp is a force of 9.806 65 N (exactly), defined as a gravitational force which acts in a vacuum on a body of a mass 1 kg at a place where there is a normal gravitational acceleration ($g_n = 9.806\ 65$ m/s^2).
[31] Sedov (1955), p. 13.
[32] Analogically, for instance, also pressure and mechanical stress, measured in pascals, as well as the module of elasticity under a strain and the modulus of volume elasticity, which are measured in newtons per square meter, have a common dimension L^1MT^{-2}.
[33] Koštál (1970), p. 13.
[34] Sedov (1955), p. 10.
[35] According to an older version of the Czechoslovak norm ČSN 01 1300, a radian is "an angle for which the ratio of the respective length of a circular arc, which is constructed from the vertex of the angle, to the radius of the arc is equal to 1" and a steradian is "a solid angle for which the ratio of the area on the surface of a sphere, cut by a cone and having its center in the vertex of the angle, to the square of the radius of the sphere, is equal to 1".
[36] Stille (1961), p. 34.
[37] Campbell (1920/1957), p. 283.
[38] Suppes (1951), p. 163f.
[39] Campbell (1920/1957), p. 348; Campbell (1928), p. 14.
[40] Cf. Campbell (1920/1957), pp. 277f, 346f; Campbell (1928), pp. 26, 101f.

[41] Brož (1967), p. 17.
[42] Campbell (1920/1957), pp. 277, 346f.
[43] In agreement with Ellis, we shall assume that this magnitude is associatively measured.
[44] Campbell (1920/1957), p. 378f.
[45] Campbell (1920/1957), p. 402.
[46] Campbell (1928), p. 101f.
[47] Smart (1959), p. 4f.
[48] Cf., e.g., Stegmüller (1970), p. 94.

CHAPTER 4

[1] Upsah (1968), p. 73.
[2] Smart (1969), pp. 7, 13. Cf. also Hempel (1952), p. 58; Krantz *et al.* (1971), p. 517.
[3] Suppes and Zinnes (1963), p. 11; Krantz *et al.* (1971), p. 143.
[4] Cf. Bunge (1967), II, p. 221f.
[5] *[There seems to be no corresponding distinction in English. We shall use the term *scale* for the Czech term *škála*, and the term *gauge* for the Czech term *stupnice* – tr.]
[6] Campbell (1920/1957), p. 321.
[7] Menger (1959), p. 105.
[8] Cf. Ellis (1966), pp. 60, 193.
[9] Cf., e.g., Ipsen (1960), p. 16.
[10] Cf. Torgerson (1958), pp. 15, 42, 35.
[11] Coombs (1952), p. 481.
[12] Cf. Pfanzagl (1959), p. 284; (1962), pp. 19f, 59f.
[13] Cf. Aczél (1948), p. 392f.
[14] Cf. for more details Berka (1974), p. 71f.
[15] Hölder (1901), p. 10.
[16] Carnap (1926), p. 28.
[17] Russell (1903/1964), pp. 171, 183.
[18] Cf., e.g., Torgerson (1958), p. 36.
[19] Pfanzagl (1962), p. 20. In fact, this axiom is a law of tautology (idempotency), which facilitates in the sentential logic a simplification of a conjunction or an alternative of the same sentential variables, or in the logic of classes a union or an intersection of classes containing the same element. In the logic of relations, the law of reflexivity is formulated as follows: $(\forall x)\, x\, R\, x$.
[20] Coombs (1953), p. 481.
[21] Cf. Bunge (1967), II, p. 222.
[22] To this differentiation there corresponds a distinction between linear and non-linear scales in the usual physical sense of the term.
[23] Cf. Stevens (1946/1960), p. 146; Coombs (1953), p. 483; Coombs *et al.* (1954), p. 137f; Ackoff *et al.* (1962), p. 191.

CHAPTER 5

[1] Cf., e.g., Stevens (1959), p. 26; Luce (1959), p. 147.

[2] Cf., e.g., Upsah (1968), p. 73; Krantz *et al.* (1971), p. 33.
[3] Cf., e.g., Hagendorf (1972), p. 68; Coombs (1952), p. 484.
[4] Ellis (1966), p. 54.
[5] E.g., Bergmann and Spence (1944), pp. 8f, 11; Suppes and Zinnes (1963), p. 16f; Ellis (1966), pp. 78, 90f.
[6] E.g., Stevens (1959), p. 45; Coombs (1952), p. 473.
[7] E.g., Bergmann and Spence (1944), p. 6; Pfanzagl *et al.* (1968), p. 31.
[8] Stevens (1951), p. 23.
[9] Ellis (1966), p. 41.
[10] Cf. Ross (1964), p. 47f; Saugstad (1965), p. 159f; Rozeboom (1966), p. 171f; Cicourel (1966), p. 119.
[11] Cf., e.g., Bunge (1967), II, pp. 197, 206.
[12] Cf. Campbell (1920/1957), pp. 301, 309; Werkmeister (1948), p. 269; Bunge (1967), II, pp. 212, 218.
[13] Cf. Campbell (1920/1957), p. 303f; Smart (1959), p. 2; Stevens (1961), p. 72; Ellis (1966), p. 152f.
[14] Ellis (1966), pp. 153, 159.
[15] Suppes and Zinnes (1963), p. 9.
[16] Stevens (1946/1960), p. 147. Cf. also Stevens (1959), p. 20.
[17] Cf., e.g., Carnap (1926), p. 15f; Kockelmans (1966), p. 115.

CHAPTER 6

[1] Kanger (1972), p. 28.
[2] Cf., e.g., Campbell (1928), pp. 14, 101.
[3] Cf., e.g., Bridgman (1927), p. 16; Dingle (1950), p. 23.
[4] Cf. Suppes and Zinnes (1963), p. 17; Pfanzagl *et al.* (1968), p. 31; Krantz *et al.* (1971), p. 1.
[5] Krantz *et al.* (1971), p. 123.
[6] Cf. Suppes and Zinnes (1963), pp. 16f, 45.
[7] *[For the corresponding pages of the quoted passages, see note 6; this is an economical practice generally used in this book – tr.]
[8] Cf. Krantz *et al.* (1971), p. 9; Suppes and Zinnes (1963), p. 16.
[9] Suppes and Zinnes (1963), p. 20f accept, moreover, a third kind of measurement – the so-called pointer measurement, i.e., a numerical assignment based on the direct readings of the scales of measuring instruments.
[10] Cf. Krantz *et al.* (1971), pp. 245f., 131f., 502 [on p. 503 there are valuable terminological comparisons; the term 'associative', e.g., is used by Ellis – tr.]; Suppes (1972), p. 53f.
[11] Kanger (1972), p. 29f.
[12] Kanger (1972), p. 12; cf. likewise Goude (1962), p. 15.
[13] Hempel (1952), p. 58f.
[14] Stegmüller (1970), pp. 2, 46, 94f.
[15] Cf., e.g., Hempel (1964), p. 112, where fundamental measurement is identified with direct measurement.
[16] Similarly, e.g., Pap (1962), p. 132f. distinguishes derived measurement on the basis of

an *empirical* function and derived measurement on the basis of a *definitional* function. Cf. also Ross (1964), p. 62.

[17] Cf., e.g., Stegmüller (1970), p. 98.

[18] Bunge (1967), II, p. 232f.

[19] Torgerson (1958), p. 21f.; cf. likewise Pfanzagl *et al.* (1968), p. 33.

[20] Ellis (1966), pp. 54f, 90f; (1967), p. 247.

[21] Some adherents of this theory regard only classificatory concepts as qualitative ones.

[22] Kutschera (1972), p. 27.

[23] Cf. Stegmüller (1970), p. 53f.

[24] Cf. Nagel (1932), p. 315; Suppes (1951), p. 164f; Hempel (1952), p. 59f; Behrend (1956), p. 346f; Ross (1964), pp. 38f, 49f; Leinfellner (1965), p. 83f; Stegmüller (1970), p. 29f; Kutschera (1972) I, p. 22f.

[25] Cf., e.g., Pfanzagl (1962), p. 15f; Suppes and Zinnes (1963), p. 34f; Krantz *et al.* (1971), p. 136f; Kanger (1972), p. 33f; Suppes (1972), p. 48f; Kutschera (1972) I, p. 23f.

[26] In the formulations of axioms (Ax), theorems (Th), definitions (Df) and correspondence rules (Ru) we use, without further explanation, the customary, familiar expression means of modern formal logic.

[27] Stegmüller (1970), p. 30 imposes on K a stronger condition: this relation must be totally reflexive; i.e., it must be valid for it, furthermore, every element of the set $\boldsymbol{E}$ must belong either to the domain or to the inverse domain of the relation K. For this reason he holds the expression $(\forall x)\ (xKx)$ to be an axiom.

[28] Cf., e.g., Tarski (1969), p. 169f.

[29] In distinction to other formulations, we introduce specific symbols for all these relations. Although the use of these symbols, usually only arithmetical ones, is more evident, there is nevertheless an easy temptation to misinterpret them, conditioned by inappropriate analogies between empirical and numerical aspects of measurement. Every reader may easily see that the relation $\boldsymbol{U}$ corresponds to the relation K, the relation $\boldsymbol{V}$ to the relation L, etc.

[30] Suppes (1972), p. 55.

[31] Neumann and Morgenstern (1947/1953), p. 627.

[32] Krantz *et al.* (1971), p. 74.

[33] Cf. Suppes and Zinnes (1963), p. 17f.

[34] Krantz *et al.* (1971), p. 8.

[35] Krantz (1972), p. 1434.

[36] Suppes (1969a), p. 2; cf. likewise Krantz *et al.* (1971), p. 8.

[37] Cf. Suppes and Zinnes (1963), pp. 19, 44f; Krantz *et al.* (1971), p. 99f.

[38] Ellis (1966), p. 79f; Campbell (1920/1957), p. 290f; Krantz *et al.* (1971), p. 87f.

[39] Krantz *et al.* (1971), p. 87.

CHAPTER 7

[1] Stevens (1946/1960), p. 142f; (1959), p. 25f; Coombs (1952), p. 472f; (1952a), p. 2f; Torgerson (1958), p. 15f; Suppes and Zinnes (1963), p. 14f; Pfanzagl *et al.* (1968), pp. 28f, 74f; Kanger (1972), p. 4. Cf. in more detail Berka (1972).

[2] Leinfellner (1965), p. 75f; Bunge (1967) II, p. 222; Ross (1954), pp. 47f, 59.

3 *[Cf. Stevens (1951), p. 26 – tr.]
4 Stevens (1946/1960), p. 146.
5 Stevens (1959), p. 32.
6 *[Cf. Stevens (1951), p. 25 – tr.]
7 Cf. Ellis (1966), p. 65f.
8 This expression can of course be expressed also as follows:

$$T = \left(100 \cdot \frac{V - V_0}{V_{100} - V_0} + 273{,}16\right) K.$$

9 Campbell (1920/1957), p. 294.
10 von Neumann and Morgenstern (1944/1953), p. 627f. Compare (6.4).
11 Cf., e.g., Friedman and Savage (1948), p. 293; Baumol (1951), p. 65; Ellsberg (1954), p. 545; Savage (1954), pp. 48, 82, 202; Luce and Raiffa (1957), p. 33; Stevens (1959), p. 45f; Adams (1964), pp. 165, 183; Fishburn (1964), p. 10; Allen (1971), p. 628.
12 Cf. in more detail Berka (1974), p. 165f.
13 Adams (1966), pp. 165 and 183.
14 Cf. Suppes (1972), p. 57f.
15 Ramsey (1931), p. 167f. Cf. also Jeffrey (1965), p. 14f; Berka (1974), p. 54f.

CHAPTER 8

1 Huntington (1902), p. 264.
2 Suppes and Zinnes (1963), p. 45.
3 Suppes (1954), p. 246; Krantz *et al.* (1971), p. 25.
4 Cf. Krantz *et al.* (1971), pp. 32f, 124.
5 Coombs (1952), p. 7.
6 Stevens (1968), p. 174. Cf. also Saugstad (1965), p. 60.
7 Cf. Berka (1974), pp. 44f, 79f.
8 Cf. Scott and Suppes (1958), p. 113; Krantz *et al.* (1971), p. 8; Suppes (1969), p. 95.
9 Stevens (1961), p. 75. Cf. likewise Ellis (1966), pp. 42, 63.
10 Russell (1903/1964), p. 160f. Cf. Adams (1965), p. 206f.
11 Lenin (1961), p. 171.
12 Luce and Raiffa (1957), p. 35; Adams (1960), p. 187; Fishburn (1970), p. 108.
13 Cf. Carnap (1966), p. 74f.
14 von Neumann and Morgenstern (1944/1953), p. 21f.
15 *[Cf. von Neumann and Morgenstern (1944/1953), p. 617 – tr.]
16 Ellis (1966), p. 160f.
17 Cf. Savage (1954), p. 32f; Fishburn (1964), p. 137f; Suppes (1969), p. 88; Adams (1960), p. 173; Leinfellner (1968), p. 198f.
18 Krantz *et al.* (1971), p. 124.
19 Brillouin (1964), p. 35.
20 Cf. e.g., Brož (1967) II, p. 32; Bunge (1967) II, p. 242f; Campbell (1920/1953), p. 471f.
21 Lamser (1966), p. 150.
22 Suppes and Zinnes (1963), pp. 64f, especially pp. 70–74.
23 *[Cf. Suppes and Zinnes (1963), p. 70 – tr.]

24 *[Cf. Suppes and Zinnes (1963), p. 71 – tr.]
25 Cf., e.g., Wilks (1961), p. 6f; Ackoff *et al* (1962), p. 200; Cicourel (1966), p. 76; Guilford (1954), p. 398; Fischer (1968), p. 31; Raatz (1968), p. 159.

CHAPTER 9

1 Cf. especially Byerly and Lazara (1973).
2 Cf. Dingle (1950), (1960), Cf. also Berka (1971).
3 Cf. Bridgman (1927), (1938), (1959), (1959a).
4 Cf., e.g., Narskii (1961), Kiseleva (1963) – Pap (1959), Hempel (1964).
5 Dingle (1950), p. 21.
6 Bridgman (1959a), p. 48.
7 Ellis (1966), pp. 38, 32, 49, 48.
8 *[In the context of the present work, the term 'magnitude' replaces Ellis' term 'quantity' – tr.]
9 Stevens (1936), p. 100.
10 Carnap (1966), pp. 99f, 105.
11 *[Cf. Carnap (1966), p. 100; the same applies to the following quotes – tr.]

BIBLIOGRAPHY

Ackoff, R. L., Gupta, S. K., and Minas, J. S.: 1962, *Scientific Method: Optimizing Applied Research Decisions*, New York: Wiley.

Aczél, J.: 1948, 'On Mean Values', *Bull. Amer. Math. Soc.* **54**, 392–400.

Adams, E. W.: 1960, 'Survey of Bernoullian Utility Theory', in H. Solomon (ed.), *Mathematical Thinking in the Measurement of Behavior*, Glencoe, Ill.: Free Press, pp. 151–268.

Adams, E. W.: 1965, 'Elements of a Theory of Inexact Measurement', *Phil. Sci.* **32**, 205–228.

Allen, R. G. D.: 1965, *Mathematical Economics*, London: Macmillan; New York: St. Martin's Press.

Baird, D. C.: 1962, *Experimentation: an Introduction to Measurement Theory and Experiment Design*, Englewood Cliffs, N.J.: Prentice-Hall.

Baumol, W.: 1951, 'The Neumann–Morgenstern Utility Index – An Ordinalist View', *J. Pol. Econ.* **59**, 61–62.

Behrend, F. A.: 1953, 'A System of Independent Axioms for Magnitudes', *J. Proc. Roy. Soc. New South Wales* 87, 27–30.

Behrend, F. A.: 1956, 'A Contribution to the Theory of Magnitudes and the Foundations of Analysis', *Math. Zeitschr.* **63**, 345–362.

Bergmann, G. and Spence, K. W.: 1944, 'The Logic of Psychophysical Measurement', *Psychol. Rev.* **51**, 1–24.

Berka, K.: 1970, 'Russell's Theory of Quantity and Magnitude', *Teorie a Metoda* **II**/2, 35–51.

Berka, K.: 1971, 'Kritický rozbor operacionalistického pojetí měření' ['A Critical Analysis of the Operationalist Doctrine of Measurement'], *Filosofický Časopis* **19**, 209–221.

Berka, K.: 1971a, 'K pojetí měření v československé sociologii' ['Toward the Conception of Measurement in Czechoslovak Sociology'], *Sociologický Časopis* **5**, 545–558.

Berka, K.: 1972, *Škály měření: Kritický rozbor pojetí škál v teorii měření a jejich funkce* [*Measurement Scales: a Critical Analysis of the Concept of Scales in Measurement Theory and of Their Function*], Prague: CSAV.

Berka, K.: 1973, 'K axiomatizaci intenzívních veličin' ['Towards the Axiomatization of Intensive Magnitudes'], *Filosofický Časopis* **21**, 252–258.

Berka, K.: 1973a, 'Lamberts Beitrag zur Messtheorie', *Organon* 9, 231–241.

Berka, K.: 1973b, 'Možnosti a meze uplatňování logicko-matematických metod ve společenských vědách' ['The Possibilities and Limits of the Application of Logico-Mathematical Methods in the Social Sciences'], in *K aktuálním otázkám naší filosofie a sociologie* [*Towards the Acute Questions of our Philosophy and Sociology*], Prague: Academia, pp. 259–264.

Berka, K.: 1974, *Teorie očekávaného užitku* [*Theory of Expected Utility*], Prague: Ústav pro filosofii a sociologii ČSAV.

Berka, K.: 1974a, 'Theoretisch-methodologische Bemerkungen zur Systemauffassung in der Wissenschaft', in G. Kröber *et al.* (eds.), *Wissenschaft und Forschung im Sozialismus: Probleme ihrer Entwicklung, Gestaltung und Analyse*, Berlin: Akademie-Verlag, pp. 222–232.

Berka, K.: 1974b, 'Ramsey's Logic of Partial Belief', *Teorie a Metoda* **VI/1**, 67–84.

Bernadette, J. A.: 1968, 'Continuity and the Theory of Measurement', *J. Phil.* **65**, 411–430.

Bernal, J. D.: 1971, *Science in History*, 4 vols., Cambridge, Mass.: M.I.T. Press.

Bernoulli, D.: 1738/1954, 'Specimen theoriae novae de mensura sortis', *Commentarii Acad. Scientiarum Imperialis Petropolitanae* **5**, 175–192; English transl.: 'Exposition of a New Theory on the Measurement of Risk', *Econometrica* **22**, 23–36.

Binko, J,: 1968, *Fyzikální a technické veličiny* [*Physical and Technical Magnitudes*], Prague: SNTL.

Bolzano, B.: 1950, *Paradoxes of the Infinite*, tr. and ed. by F. Příhonský, London: Routledge and Kegan Paul.

Bridgman, P. W.: 1927, *The Logic of Modern Physics*, New York: Macmillan.

Bridgman, P. W.: 1931, *Dimensional Analysis*, New Haven, Conn.: Yale University Press.

Bridgman, P. W.: 1938, 'Operational Analysis Research', *Phil. Sci.* **5**, 114–131.

Bridgman, P. W.: 1959, 'How Much Rigor is Possible in Physics?', in L. Henkin *et al.* (eds.), *The Axiomatic Method*, Amsterdam, pp. 225–237.

Bridgman, P. W.: 1959a, *The Way Things Are*, Cambridge, Mass.: Harvard University Press.

Brillouin, L.: 1964, *Scientific Uncertainty and Information*, New York: Academic Press.

Brož, J. *et al.*: 1967, *Základy fyzikálních měření* [*The Elements of Physical Measurements*], Prague: SPN.

Buckingham, E. B.: 1914, 'On Physically Similar Systems: Illustrations of the Use of Dimensional Equations', *Phys. Rev.* **4**, 345–376.

Buckingham, E. B.: 1924, 'Dimensional Analysis', *Phil. Mag.*, Ser. 6, **48**, 141–145.

Bunge, M.: 1967, *Scientific Research*, Vols. I and II, Berlin: Springer-Verlag.

Byerly, H. C. and Lazara, V. A.: 1973, 'Realist Foundations of Measurement', *Phil. Sci.* **40**, 10–28.

Campbell, N. R.: 1920/1957, *Physics: The Elements*, Cambridge, Engl.: Cambridge University Press; repr. as *Foundations of Science: The Philosophy of Theory and Experiment*, New York: Dover, 1957.

Campbell, N. R.: 1928, *An Account of the Principles of Measurement and Calculation*, New York: Longmans, Green.

Campbell, N. R.: 1940, 'Notes on Physical Measurement', in *Final Report, The Advancement of Science* I, 2, pp. 340–342.

Carnap, R.: 1926, *Physikalische Begriffsbildung*, Karlsruhe: Braun.

Carnap, R.: 1966, *Philosophical Foundations of Physics*, New York: Basic Books.

Causey, R. L.: 1969, 'Derived Measurement, Dimensions, and Dimensional Analysis', *Phil. Sci.* **36**, 252–270.

Caws, P.: 1959, 'Definition and Measurement in Physics', in C. W. Churchman and P. Ratoosh (1959), pp. 3–17.

Chernoff, H. and Moses, L. E.: 1959, *Elementary Decision Theory*, New York, London, Sydney (5th ed., 1967): Wiley.

Churchman, C. W. and Ratoosh, P. (eds.): 1969, *Measurement; Definitions and Theories*, New York: Wiley.

Cicourel, A. V.: 1966, *Method and Measurement in Sociology*, New York: Free Press of Glencoe.

Cohen, M. R. and Nagel, E.: 1934, *An Introduction to Logic and Scientific Method*, New York: Harcourt, Brace.

Coombs, C. H.: 1952, 'A Theory of Psychological Scaling', *Eng. Res. Instit. Bull.*, No. 34, Ann Arbor, Mich.

Coombs, C. H.: 1953, 'Theory and Methods of Social Measurement', in L. Festinger and O. Katz (eds.), *Research Methods in the Behavioral Sciences*, New York: Holt, Rinehart and Winston, pp. 471–535.

Coombs, C. H.: 1965, 'Psychological Scaling Without a Unit of Measurement', in R. D. Luce *et al.* (eds.), *Readings in Mathematical Psychology*, vol. II, New York, London, Sydney: Wiley, pp. 451–464.

Coombs, C. H., Raiffa, H., and Thrall, R. M.: 1954, 'Some Views on Mathematical Models and Measurement Theory', *Psychol. Rev.* **61**, 132–144.

Čížek, F.: 1974, *Teorie a empirie* [*Theory and the Empirical*], Prague: Svoboda.

Danto, A. and Morgenbesser, S. (eds.): 1960, *Philosophy of Science*, Cleveland: World Pub. Co.

Dingle, H.: 1942, 'On the Dimension of Physical Magnitude', *Phil. Mag.* **33**, 321–344.

Dingle, H.: 1950, 'A Theory of Measurement', *Brit. J. Phil. Sci.* **1**, 5–26.

Dingle, H.: 1960, Review of C. W. Churchman and P. Ratoosh (eds.), 'Measurement: Definitions and Theories', *Scientific American* **202**, 6, 189–192.

Ellis, B.: 1966, *Basic Concepts of Measurement*, London (2nd ed. 1968).

Ellsberg, D.: 1954, 'Classic and Current Notions of "Measurable Utility" ', *Econ. J.* **59**, 528–556.

Engels, F.: 1964, *Dialectics of Nature*, 3rd rev. ed., Moscow.

Essler, W. K.: 1971, *Wissenschaftstheorie*, vol. II, Freiburg: K. Alber.

Eysenck, H. J. (ed.): 1976, *The Measurement of Personality*, Baltimore: University Park Press.

Farrington, B.: 1969, *Science in Antiquity*, 2nd ed., London, Oxford: Oxford University Press.

Feather, N.: 1959, *An Introduction to the Physics of Mass, Length and Time*, Edinburgh: University Press.

Fechner, G. T.: 1860, *Elemente der Psychophysik*, Leipzig: Breitkopf and Härtel.

Filkorn, V.: 1953, *Predheglovská logika* [*Pre-Hegelian Logic*], Bratislava: Vydavatel'stvo Slovenskej akadémie vied.

Filkorn, V.: 1963, *Pre-dialectical Logic*, Bratislava: Publishing House of the Slovak Academy of Sciences.

Fischer, G. H.: 1968, 'Neuere Entwicklungen in der psychologischen Testtheorie', in G. H. Fischer (ed.), *Psychologische Testtheorie*, Bern, Stuttgart: Huber, pp. 13–158.

Fishburn, P. C.: 1964, *Decision and Value Theory*, New York: Wiley.

Fishburn, P. C.: 1968, 'Utility Theory', *Manag. Sci.* **14**, 335–378.

Fishburn, P. C.: 1970, *Utility Theory for Decision Making*, New York: Wiley.

Flaschner, L.: 1952, 'Zur ontologischen Begründung der Massbegriffe und Gesetze in Mechanik und Elektrizitätslehre', *Phil. Naturalis* **2**, 137–177.

Focken, C. M.: 1953, *Dimensional Methods and Their Applications*, London: Edward Arnold.

Friedman, M. and Savage, L. J.: 1948, 'The Utility Analysis of Choices Involving Risk', *J. Pol. Econ.* **56**, 279–304.

Gnoseologicheskie aspekty izmerenii [*Gnosiological Aspects of Measurements*] : 1968, Kiev: Naukova dumka.

Goude, G.: 1962, *On Fundamental Measurement in Psychology*, Stockholm, Göteborg, Uppsala: Almqvist and Wiksell.

Guilford, J. P.: 1954, *Psychometric Methods*, 2nd ed., New York: McGraw-Hill.

Hagendorf, H.: 1972, 'Über die Rolle der Skalierung im Problemlösen', in F. Klix *et al.* (eds.), *Analyse und Synthese von Problemlösungsprozessen, Kybernetik-Forschung*, Vol. 2, Berlin: Deutscher Verlag der Wissenschaften, pp. 67–89.

Hempel, C. G.: 1952, 'Fundamentals of Concept Formation in Empirical Science', *Intern. Enc. of Unified Science* **II**/7, Chicago (1958 ed.).

Hempel, C. G.: 1960, 'Operationism, Observation, and Scientific Terms', in Danto and Morgenbesser (1964), pp. 101–120.

Hempel, C. G.: 1966, *Philosophy of Natural Science*, Englewood Cliffs, N.J.: Prentice-Hall.

Hilbert, D.: 1900, 'Über den Zahlbegriff', *Jahresber. Deutsch. Math. Ver.* **8**, 180–194.

Hilbert, D.: 1930, *Grundlagen der Geometrie*, 7th ed. Leipzig, Berlin: Teubner.

Hölder, O.: 1901, 'Die Axiome der Quantität und die Lehre vom Mass', *Ber. Kgl. Sachsis. Ges. Wiss.*, Leipzig, Math.-Phys. Classe 53, pp. 1–64.

Hofmann, K. H.: 1963, 'Zur mathematischen Theorie des Messens', *Rozpr. Mat.* **32**, 1–31.

Huntington, E. V.: 1902, 'A Complete Set of Postulates for the Theory of Absolute Continuous Magnitude', *Trans. Amer. Math. Soc.* **3**, 264–279.

Ipsen, D. C.: 1960, *Units, Dimensions and Dimensionless Numbers*, New York: McGraw-Hill.

Jeffrey, R. C.: 1965, *The Logic of Decision*, New York: McGraw-Hill.

Jeffreys, H.: 1943, 'Units and Dimensions', *Phil. Mag.* 7, Ser. 34, 837–842.

Kanger, S.: 1972, 'Measurement: An Essay in Philosophy of Science', *Theoria* **38**, 1–44.

Kiseleva, N. A.: 1963, 'O sovremennom sostojanii operacionalizma P. Bridgmana', in *Filosofiia marxizma i neopositivizm* [*The Philosophy of Marxism and Neopositivism*], Moscow: Izd-vo Moskovskogo Univ.

Klein, H. A.: 1974, *The World of Measurement Measurements*, New York: Simon and Schuster.

Kockelmans, J. J.: 1966, *Phenomenology and Physical Science: An Introduction to the Philosophy of Physical Sciences*, Pittsburgh, Pa.: Duquesne University Press.

Körner, S.: 1966, *Experience and Theory*, New York: Humanities Press.

Koštál, K.: 1970, *Sbírka fyzikálních vzorců a poučék* [*A Collection of Physical Patterns and Theorems*], Prague: SNTL, t. Mír.

Krantz, D. H.: 1972, 'Measurement Structures and Psychological Laws', *Science* **175**, 1427–1435.

Krantz, D. H., Luce, R. D., Suppes, P., and Tversky, A. A.: 1971, *Foundations of Measurement*, Vol. I, New York, London: Academic Press.

Kyburg, H. Jr.: 1969, 'Comments and Criticisms: Measurement and Mathematics', *J. Phil.* 66, 29–42.

Lambert, J. H.: 1771, *Anlage zur Architectonic; oder Theorie des Einfachen und des Erstens in der philosophischen und mathematischen Erkenntnis*, Riga: Hartknoch, Vol. 2, Pars. 679–923, pp. 301–560.

Lamser, V.: 1966, *Základy sociologického výzkumu* [*The Foundations of Sociological Inquiry*], Prague: Svoboda.

Lazarsfeld, P.: 1960, 'Evidence and Inference in Social Research', in D. Lerner (ed.), *Evidence and Inference*, Chicago: Free Press of Glencoe; essays in this volume were first published in *Daedalus*, 1959.

Leinfellner, W.: 1965, *Struktur und Aufbau wissenschaftlicher Theorien: Eine wissenschaftstheoretisch-philosophische Untersuchung*, Vienna, Würzburg: Physica-Verlag.

Leinfellner, W.: 1968, 'Generalization of Classical Decision Theory', in K. H. Borch and J. Mossin (eds.), *Risk and Uncertainty*, New York: St. Martin's, pp. 196–210.

Lenin, V. I.: 1961, *Philosophical Notebooks*, Vol. 38 of *Collected Works*, London: Lawrence and Wishart.

Levin, M. E.: 1971, 'Length Relativity', *J. Phil.* **68**, 164–174.

Lindsay, R. B.: 1938, 'The Future of Theoretical Physics', *Phil. Sci.* **5**, 452–471.

Luce, R. D.: 1959, 'A Probabilitic Theory of Utility and Its Relationship to Fechnerian Scaling', in Churchman and Ratoosh (1959), pp. 144–159.

Luce, R. D.: 1971, 'Similar Systems and Dimensionally Invariant Laws', *Phil. Sci.* **38**, 157–169.

Luce, R. D. and Raiffa, H.: 1957, *Games and Decisions: Introduction and Critical Survey*, New York: Wiley.

Margenau, H.: 1950, *The Nature of Physical Reality: A Philosophy of Modern Physics*, New York: McGraw-Hill.

Marinitchev, E. A.: 1971, 'O poniatiach kolichestva, velichiny i chisla' ['On the Concepts of Quantity, Magnitude and Number'], in *Aktual'nye problemy dialekticheskoi logiki* [*The Acute Problems of Dialectical Logic*], Alma-Ata: Izd-vo Nauka Kazakhskoi SSR, pp. 439–444.

Menger, K.: 1959, 'Mensuration and Other Mathematical Connections of Observable Material', in Churchman and Ratoosh (1959), pp. 97–128.

Metodologicheskie problemy teorii izmerenii [*Methodological Problems in the Theory of Measurement*]: 1966, Kiev: Naukova dumka.

Nagel, E.: 1932, 'Measurement', *Erkenntnis* **2**, 313–333; repr. in Danto and Morgenbesser (1960), pp. 121–140.

Narskii, I. S.: 1961, *Sovremennyi pozitivizm* [*Contemporary Positivism*], Moscow: Akad. nauk.

O'Rahilly, A.: 1938, *Electromagnetics: A Discussion of Fundamentals*, London: Longmans, Green.

Osipov, G. V. (ed.): 1966, *Kolichestvennye metody v sociologii* [*Quantitative Methods in Sociology*], Moscow: Nauka.

Page, C. H. and Vigoureux, P. (eds.): 1975, *The International Bureau of Weights and Measures 1875–1975*, U.S. Department of Commerce.

Pap, A.: 1959, 'Are Physical Magnitudes Operationally Definable?', in Churchman and Ratoosh (1959), pp. 177–191.

Pap, A.: 1962, *An Introduction to the Philosophy of Science*, New York: Free Press of Glencoe.

Peak, H.: 1953, 'Problems of Objective Observation', in L. Festinger and D. Katz (eds.), *Research Methods in Behavioral Science*, New York: Holt, Rinehart and Winston, pp. 243–299.

Pfanzagl, J.: 1959, 'A General Theory of Measurement – Applications to Utility', *Naval. Res. Logistics Quart.* **6**, 283–294; repr. in R. D. Luce *et al.* (eds.), *Readings in Mathematical Psychology*, Vol. II, New York, London, Sydney: Wiley, 1965, pp. 492–502.

Pfanzagl, J., Bauman, V., and Huber, H.: 1968, *Theory of Measurement*, Würzburg, Vienna: Physica-Verlag.

Przełęcki, M.: 1969, *The Logic of Empirical Theories*, London: Routledge and Kegan Paul.

Raatz, V.: 1968, 'Neuere Ansätze zur Theorie der Reliabilität', in G. H. Fischer (ed.), *Psychologische Testtheorie*, Bern, Stuttgart: Huber, pp. 159–176.

Radnitzky, G.: 1973, 'Toward a Theory of Research which is neither Logical Reconstruction nor Psychology or Sociology of Science', *Teorie a Metoda* V/2, 25–54.

Ramsey, F. P.: 1931, 'Truth and Probability', in *The Foundations of Mathematics and Other Logical Essays*, New York: Harcourt, Brace, pp. 156–198; repr. in H. E. Kyburg, Jr. and H. E. Smokler (eds.), *Studies in Subjective Probability*, New York: Wiley, 1964, pp. 61–92.

Ross, S.: 1964, *Logical Foundations of Psychological Measurement: A Study in the Philosophy of Science*, Copenhagen: Munksgaard.

Rozeboom, W. W.: 1966, 'Scaling Theory and the Nature of Measurement', *Synthese* **16**, 170–233.

Russell, B.: 1903/1964, *Principles of Mathematics*, Cambridge: Cambridge University Press; 8th ed., London: Allen and Unwin, 1964.

Saugstad, P.: 1965, *An Inquiry into the Foundations of Psychology*, Oslo: Universitetsforlaget; New York: Bedminster Press.

Savage, C. W.: 1970, *The Measurement of Sensation*, Berkeley: University of California Press.

Savage, L. J.: 1954, *The Foundations of Statistics*, New York: Wiley.

Scott, D. and Suppes, P.: 1958, 'Foundational Aspects of Theories of Measurement', *J. Symb. Logic* **23**, 113–128.

Sedov, L. I.: 1955, *Metody podobnosti a rozměrovosti v mechanice* [*Methods of Similarity and Dimensionality in Mechanics*], Prague.

Sena, L. A.: 1953, *Fyzikální jednotky* [*Physical Units*], Prague.

Schleichert, H.: 1964, 'Zur Erkenntnislogik des Messens', *Arch. Phil.* **12**, 304–327.

Schleichert, H.: 1966, *Elemente der physikalischen Semantik*, Vienna, Munich: Oldenbourg.

Smart, J. J. C.: 1959, 'Measurement', *Austral. J. Phil.* **37**, 1–13.

Stahl, W. R.: 1962, 'Similarity and Dimensional Methods in Biology', *Science* **137**, 205–212.

Staríček, I.: 1968, 'Proces merania z hľadiska metronomiky' ['The Process of Measurement from the Standpoint of Metronomics'], *Acta Metronomica*, ÚTM–SAV, 4.

Staríček, I.: 1971, 'Metronika meracieho systému' ['Metronics of a Measuring System'], *Acta Metronomica*, ÚTM–SAV, 7.

Staríček, I.: 1972, 'Metronomische Analyse physikalischer und technischer Messungen', *Acta Metronomica*, ÚTM–SAV, 8.

Staríček, I.: 1977, *Úvod do metronomiky* [*Introduction to Metronomics*], Bratislava.

Stegmüller, W.: 1970, *Probleme und Resultate der Wissenschaftstheorie und Analytischen Philosophie*, Vol. 2: *Theorie und Erfahrung*, 2 vols. Berlin, Heidelberg, New York: Springer-Verlag.

Stevens, S. S.: 1936, 'Psychology: The Propaedeutic Science', *Phil. Sci.* **3**, 90–103.

Stevens, S. S.: 1964/1960, 'On the Theory of Scales of Measurement', *Science* **103**, 677–680; repr. in Danto and Morgenbesser (1960), pp. 141–149.

Stevens, S. S.: 1951, 'Mathematics, Measurement and Psychophysics', in S. S. Stevens (ed.), *Handbook of Experimental Psychology*, New York: Wiley, pp. 1–49.

Stevens, S. S.: 1959, 'Measurement, Psychophysics, and Utility', in Churchman and Ratoosh (1959), pp. 18–63.

Stevens, S. S.: 1968, 'Ratio Scales of Opinion', in D. K. Whitla (ed.), *Handbook of Measurement and Assessment in Behavioral Sciences*, Reading, Mass.: Addison-Wesley, pp. 171–199.

Stille, U.: 1955, *Messen und Rechnen in der Physik*, Braunschweig: F. Vieweg.

Stolz, O. and Gmeiner, J. A.: 1900, *Theoretische Arithmetik*, Vol. I, Leipzig: Teubner.

Struik, D. J.: 1967, *A Concise History of Mathematics*, 3rd rev. ed., New York: Dover.

Suppes, P.: 1951, 'A Set of Independent Axioms for Extensive Quantities', *Portugal. Math.* **10**, 163–172.

Suppes, P.: 1954, 'Some Remarks on Problems and Methods in the Philosophy of Science', *Phil. Sci.* **21**, 242–248.

Suppes, P.: 1959, 'Measurement, Empirical Meaningfulness, and Three-valued Logic', in Churchman and Ratoosh (1959), pp. 129–143.

Suppes, P.: 1969, *Studies in the Methodology and Foundations of Science: Selected Papers from 1951 to 1969*, Dordrecht, Holland: D. Reidel.

Suppes, P.: 1969a, 'Measurement: Problems of Theory and Applications', *Stanford Technical Report*, Psychological Series, Rep. No. 147.

Suppes, P.: 1972, 'Finite Equal-Interval Measurement Structure', *Theoria* 38, 45–63.

Suppes, P. and Zinnes, J. L.: 1963, 'Basic Measurement Theory', in R. Luce *et al.* (eds.), *Handbook of Mathematical Psychology*, New York: Wiley, pp. 1–76.

Shtoff, V. A., Novikov, A. J., and Jadov, V. A. (eds.): 1970, *Problemy metodologii social'nogo issledovania* [*Problems of Methodology of Social Investigation*], Leningrad: Izd. Leningrad. un-ta.

Shubkin, V. N.: 1970, *Sociologicheskie opyty* [*Sociological Experiences*], Moscow: Mysl'.

Tarski, A.: 1954–1955, 'Contributions to the Theory of Models I–III', *Indagationes Math.* **16**, 572–581.

Tarski, A.: 1965, *Introduction to Logic and to the Methodology of Deductive Sciences*, 3rd. ed. New York: Oxford University Press.

Thiel, R.: 1967, *Quantität oder Begriff?*, Berlin: Deutscher Verlag der Wissenschaften.

Torgerson, W. S.: 1958, *Theory and Methods of Scaling*, New York: Wiley.

Ujemov, A. I.: 1963, *Veshchi, svojstva, otnoshenia* [*Things, Properties, Relations*], Moscow: Izd-vo Akademii Nauk SSSR.

Ujemov, A. I.: 1970, *Sistemy i sistemnye issledovania* [*Systems and System Investigations*], in I. V. Blauberg *et al.* (eds.), *Problemy metodologii sistemnogo issledovania* [*Problems of Methodology of System Investigation*], Moscow: Mysl'.

Upsah, H. S.: 1968, 'Attitude Measurement', in H. M. Blalock and A. B. Blalock (eds.), *Methodology in Social Research*, New York: McGraw-Hill, pp. 60–111.

von Alberti, H. J.: 1957, *Mass und Gewicht*, Berlin: Akademie-Verlag.

von Helmholtz, H.: 1887/1971, 'Zählen und Messen erkenntniss-theoretisch betrachtet', in *Philosophische Aufsätze Eduard Zeller gewidmet*, Leipzig: Fues's Verlag, pp. 17–52; repr. in *Wissenschaftliche Abhandlungen*, Vol. 3, Leipzig: Barth, 1895, pp. 356–391. Engl. tr. in R. Kahl (ed.), *Selected Writings of H. von Helmholtz*, Middletown, Conn.: Wesleyan University Press, 1971, pp. 437–465.

von Kutschera, F.: 1972, *Wissenschaftstheorie*, Vol. I, Munich: Fink.
von Neumann, J. and Morgenstern, O.: 1944/1953, *Theory of Games and Economic Behavior*, Princeton, N.J.: Princeton University Press (3rd ed., 1953).
Wartofsky, M. W.: 1968, *Conceptual Foundations of Scientific Thought*, New York: Macmillan.
Werkmeister, W. H.: 1948, *The Basis and Structure of Knowledge*, New York: Harper.
Weyl, H.: 1927, *Philosophie der Mathematik und Naturwissenschaft*, Munich, Berlin: Oldenbourg; *Philosophy of Mathematics and Natural Science*, tr. O. Helmer, Princeton, N. J. Princeton University Press, 1949.
Wiener, N.: 1921, 'A New Theory of Measurement: A Study in the Logic of Mathematics', *Proc. London Math. Soc.* **19**, 181–205.
Wiggins, J. A.: 1968, 'Hypothesis, Validity and Experimental Laboratory Methods', in H. M. Blalock and A. B. Blalock (eds.), *Methodology in Social Research*, New York: McGraw-Hill, pp. 390–427.
Wilks, S. S.: 1961, 'Some Aspects of Quantification in Science', in H. Woolf (ed.), *Quantification*, Indianapolis: Bobbs-Merrill, pp. 5–12.

INDEX OF PERSONAL NAMES

Compiled by Susan Riska

INDEX OF SUBJECTS

Compiled by Augustin Riska

GPSR Compliance

The European Union's (EU) General Product Safety Regulation (GPSR) is a set of rules that requires consumer products to be safe and our obligations to ensure this.

If you have any concerns about our products, you can contact us on ProductSafety@springernature.com

In case Publisher is established outside the EU, the EU authorized representative is:

Springer Nature Customer Service Center GmbH
Europaplatz 3
69115 Heidelberg, Germany

Batch number: 10761599

Printed by Printforce, the Netherlands